Is Your Math Ready For Calculus?

Walter J. Gleason
Bridgewater State College

WCB **Wm. C. Brown Publishers**
Dubuque, Iowa•Melbourne, Australia•Oxford, England

Book Team

Editor *Earl McPeek*
Developmental Editor *Theresa Grutz*
Production Editor *Audrey A. Reiter*

Wm. C. Brown Publishers
A Division of Wm. C. Brown Communications, Inc.

Vice President and General Manager *Beverly Kolz*
Vice President, Director of Sales and Marketing *John W. Calhoun*
Marketing Manager *Julie Keck*
Advertising Manager *Janelle Keeffer*
Director of Production *Colleen A. Yonda*
Publishing Services Manager *Karen J. Slaght*

Wm. C. Brown Communications, Inc.

President and Chief Executive Officer *G. Franklin Lewis*
Corporate Vice President, President of WCB Manufacturing *Roger Meyer*
Vice President and Chief Financial Officer *Robert Chesterman*

Cover design by John R. Rokusek

A Times Mirror Company

Library of Congress Catalog Card Number: 93–70764

ISBN 0–697–21495–8

Printed in the United States of America by Wm. C. Brown Communications, Inc., 2460 Kerper Boulevard, Dubuque, IA 52001

10 9 8 7 6 5 4 3 2

TABLE OF CONTENTS

PART I: A Diagnostic Test and A Pretest

Sections 1 to 14 inclusive are for all students who are enrolling in a calculus course.

The remaining sections are for students who are enrolling in calculus and who plan to major in mathematics or physics.

PART II: A Review of Basic Concepts

The remaining sections are for students who are enrolling in
calculus and who plan to major in mathematics or physics.

INTRODUCTION FOR NONMATHEMATICS MAJORS

Part I of this booklet is a **diagnostic test** designed to see if you are prepared for the mathematics that will appear in a calculus course. Do all of the odd-numbered exercises from 1 to 459 inclusive. Correct your answers by comparing them to the odd-numbered answers at the end of Part I. If you have scored 184 or more correct answers (80% of 230 exercises), enter the calculus course with the confidence that you are mathematically prepared.

If you have scored less than 184 correct answers, begin a problem-by-problem study of each exercise that was incorrect. Here, Part II of this booklet will help. Part II is a **review of basic concepts** and it includes many workable examples. The identical subheadings of Part I and Part II will allow you to focus upon those topics that bother you.

When you feel that you understand your mistakes, return to Part I, working all of the even-numbered exercises from 2 to 460, inclusive. Correct your answers by comparing them to the even-numbered answers at the end of Part I. If you have scored 184 or more correct answers, enter the calculus course with confidence.

If you have failed to achieve 184 correct answers, repeat the cycle problem-by-problem, using Part II to strengthen your skills. When you feel that you understand your mistakes, redo the odd-numbered exercises, but now you should be able to achieve 207 correct answers (90% of 230 exercises).

You may use this instrument as a **pretest** by doing the 50 exercises that are circled $\bigcirc$, and correcting your answers. A score of 40 or more correct answers (80% of 50 exercises) indicates that you are prepared for the calculus course.

ACKNOWLEDGEMENTS

Finally, I must thank Patricia Shea who is responsible for the preparation of this booklet, and Gary Gleason for proofreading the manuscript.

Please feel free to write to: Professor Walter Gleason, c/o Department of Mathematics and Computer Science, Bridgewater State College, Bridgewater, MA 02325.

INTRODUCTION FOR MATHEMATICS MAJORS

Part I of this booklet is a **diagnostic test** designed to see if you are prepared for the mathematics that will appear in a calculus course. Do all of the odd-numbered exercises from 1 to 560 inclusive. Correct your answers by comparing them to the odd-numbered answers at the end of Part I. If you have scored 224 or more correct answers (80% of 280 exercises), enter the calculus course with the confidence that you are mathematically prepared.

If you have scored less than 224 correct answers, begin a problem-by-problem study of each exercise that was incorrect. Here, Part II of this booklet will help. Part II is a **review of basic concepts** and it includes many workable examples. The identical subheadings of Part I and Part II will allow you to focus upon those topics that bother you.

When you feel that you understand your mistakes return to Part I and do all of the even-numbered exercises from 2 to 560 inclusive. Correct your answers by comparing them to the even-numbered answers at the end of Part I. If you have scored 224 or more correct answers, enter the calculus course with confidence.

If you have failed to achieve 224 correct answers, repeat the cycle problem-by-problem using Part II to strengthen your skills. When you feel that you understand your mistakes, redo the odd-numbered exercises, but now you should be able to achieve 252 correct answers (90% of 280 exercises).

You may use this instrument as a **pretest** by doing the 50 exercises that are blocked □, and correcting your answers. A score of 40 or more correct answers (80% of 50 exercises) indicates that you are prepared for the calculus course.

PART I

A DIAGNOSTIC TEST AND A PRETEST

1.1 FRACTIONS

Do each of the following computations.

1. $\dfrac{3}{6} + \dfrac{2}{6}$

2. $\dfrac{2}{9} + \dfrac{7}{9}$

3. $\dfrac{3}{4} + \dfrac{5}{8}$

4. $\dfrac{1}{9} + \dfrac{1}{4}$

5. $\dfrac{5}{14} + \dfrac{3}{7}$

6. $\dfrac{7}{16} + \dfrac{3}{4}$

7. $\dfrac{5}{16} + \dfrac{3}{5}$

8. $\dfrac{1}{4} + \dfrac{5}{14}$

9. $\dfrac{5}{12} + \dfrac{7}{18}$

10. $\dfrac{3}{7} + \dfrac{5}{16}$

11. $\dfrac{4}{7} - \dfrac{2}{7}$

12. $\dfrac{13}{25} - \dfrac{7}{25}$

13. $\dfrac{3}{4} - \dfrac{1}{2}$

14. $\dfrac{9}{10} - \dfrac{3}{20}$

15. $\dfrac{11}{12} - \dfrac{5}{8}$

16. $\dfrac{5}{6} - \dfrac{2}{7}$

17. $10 - \dfrac{7}{8}$

18. $\dfrac{5}{8} - \dfrac{2}{7}$

19. $\dfrac{9}{2} - \dfrac{3}{5}$

20. $\dfrac{3}{7} - \dfrac{1}{9}$

21. $\dfrac{3}{5} \times \dfrac{10}{11}$

22. $\dfrac{1}{7} \times \dfrac{7}{8}$

23. $\dfrac{15}{20} \times \dfrac{10}{12}$

24. $\dfrac{2}{3} \times 10$

25. $\dfrac{16}{18} \times \dfrac{2}{3}$

26. $\dfrac{9}{2} \times 1\dfrac{2}{15}$

27. $\dfrac{3}{4} \times 1\dfrac{1}{3}$

28. $\dfrac{8}{40} \times \dfrac{32}{16}$

29. $8 \times \dfrac{3}{4}$

30. $\dfrac{51}{100} \times 2\dfrac{16}{17}$

31. $\dfrac{7}{9} \div \dfrac{7}{9}$

32. $\dfrac{5}{6} \div \dfrac{1}{6}$

33. $\dfrac{9}{10} \div \dfrac{1}{10}$

34. $\dfrac{1}{8} \div \dfrac{1}{2}$

35. $\dfrac{4}{9} \div \dfrac{1}{9}$

36. $\dfrac{7}{4} \div 5\dfrac{3}{5}$

37. $\dfrac{1}{10} \div 10$

38. $\dfrac{28}{45} \div \dfrac{7}{18}$

39. $\dfrac{7}{4} \div 4$

40. $\dfrac{25}{30} \div \dfrac{5}{10}$

1.2 DECIMALS

Do each of the following computations.

41.
```
   0.5
 + 0.7
```

42.
```
   5.5
 + 5.5
```

43.
```
   8.6
 + 9.9
```

44.
```
  11.6
 + 8.8
```

45.
```
  20.75
 + 37.64
```

46.
```
  210.17
 + 66.77
```

(47.)
```
   0.7
   0.8
 + 1.4
```

48.
```
  567.74
 + 102.64
```

49.
```
   0.056
 + 0.007
```

50.
```
  55.162
   7.04
 + 9.002
```

51.
```
   1.8
 - 1.2
```

52.
```
   9.7
 - 7.2
```

53.
```
   2.1
 - 0.4
```

54.
```
  10.5
 - 3.9
```

55.
```
  598.01
 - 55.63
```

56.
```
  291.44
 - 186.56
```

57.
```
  100
 - 19,07
```

58.
```
  72.09
 - 17.60
```

(59.)
```
  19.2
 - 0.63
```

60.
```
  55
 - 0.75
```

61.
```
   2.4
 × 10
```

62.
```
   5.2
 × 8
```

63.
```
   2.8
 × 0.3
```

(64.)
```
   4.7
 × 4.7
```

65.
```
  12.7
 × 0.4
```

66.
```
  35.6
 × 9.1
```

67.
```
  27
 × 0.4
```

68.
```
    19.5
 × 1,000
```

69.
```
   2.20
 × 2.02
```

70.
```
  64.17
 × 1.2
```

71. $2\overline{)8.6}$

72. $10\overline{)19.8}$

73. $18\overline{)129.6}$

74. $8\overline{)38.4}$

75. $8\overline{)33.12}$

76. $3.8\overline{)17.898}$

77. $1.1\overline{)0.242}$

(78.) $3.8\overline{)15.96}$

79. $38.54 \div 4.7$

80. $14.03 \div 2.3$

2

1.3 RATIO AND PROPORTION

In exercises 81 to 86, state the given ratio in simplest form.

81. 4:8
82. 5:15

83. 36:16
84. 20:25

85. 4 meters: 16 meters
86. $0.35: $0.85

87. For the proportion $\frac{3}{4} = \frac{15}{20}$ state the means.

88. For the proportion $\frac{2}{7} = \frac{6}{21}$ state the extremes.

In problems 89 to 96, solve the proportion for the unknown.

89. $\frac{7}{8} = \frac{x}{40}$

90. $\frac{9}{2} = \frac{63}{y}$

91. $\frac{11}{x} = \frac{132}{24}$

92. $\frac{2}{3} = \frac{10}{n}$

93. $\frac{5}{13} = \frac{x}{65}$

94. $\frac{9}{8} = \frac{y}{32}$

95. $\frac{7}{x} = \frac{1}{9}$

96. $\frac{12}{5} = \frac{60}{x}$

97. If one inch equals 2 1/2 centimeters, how many inches equal 100 centimeters?

98. If 90 feet of wire weighs 18 pounds, what will 110 feet of the same wire weigh?

99. If $\frac{70}{x} = \frac{5}{7}$, find the value of x.

100. If $\frac{x}{48} = \frac{5}{3}$, find the value of x.

1.4 EXPONENTS

In problems 101 to 124, evaluate the given expression.

101. 5^0

102. 8^{-1}

103. 5^{-3}

104. 2^{-5}

105. 2^7

106. $(0.4)^3$

107. $(\frac{3}{4})^2$

108. $(\frac{2}{5})^{-3}$

109. $(\frac{1}{3})^2$

110. $(\frac{3}{2})^{-3}$ 111. $3^4 \cdot 3^2$ 112. $10 \cdot 10^4$

113. $(-2)^5 \cdot (-2)^3$ 114. $(x^4)^3$ 115. $(2^4)^5$

116. $(3^3)^3$ 117. $(10^2)^{-1}$ 118. $(4^{-2})^{-2}$

119. $(8^{-1})^4$ 120. $\frac{4^7}{4^2}$ 121. $\frac{6^{11}}{6^7}$

122. $\frac{x^7}{x^4}$ 123. $\frac{2^4}{2^6}$ 124. $\frac{8^5}{8^9}$

1.5   FACTORING

If possible, factor each of the following expressions.

125. $3x^2 + 6xy + 9x^3$

126. $ax^2 - 3ax + 5a^2$

127. $25y^2 - 9$

128. $100x^2 - 1$

129. $b^2 + 81$

130. $9k^2 + 4$

131. $18x^2 - 98y^2$

132. $7x^2 - 28y^2$

133. $x^3 + y^3$

134. $w^3 + 1$

135. $t^3 - 27$

136. $a^4 - b^4$

137. $x^2 + 2x + 9$

138. $y^2 - 3y + 8$

139. $x^2 + 15 - 8x$

140. $b^2 + 21 + 10b$

141. $35k^2 + 13k - 4$

4

142. $6p^2 + 13p + 5$

143. $4k^2 + 8k - 21$

144. $6x^2 + 5x + 4$

145. $2y^2 - 8y + 5$

146. $2t^2 - 3t + 1$

147. $40x^2 + 39x - 40$

148. $56y^2 + 15y - 56$

149. $12x^2 - 13x - 14$

150. $16a^2 + 56a + 49$

151. $12x^2 - 29x + 14$

152. $10r^2 + r - 2$

153. $-10y + 4y^2 + 6$

154. $9 + 21x + 6x^2$

155. $5a^4 - 5b^4$

156. $7x^4 - 7y^4$

1.6 LINEAR EQUATIONS

Solve each of the following equations for the indicated variable.

157. $5 - x = 3 + x$

158. $2 - 2x = -1 + x$

159. $7x - 5 + 2x = 19$

160. $5x + 3 - 2x = 18$

161. $8x - 13x = 2x + 56$

162. $12x - 7x - 4x + 9 = 19$

163. $2(3n-5) = -15$

164. $3(2b-4) = 18$

165. $x - (88-x) = 36$

166. $-7b + 4(2b-3) = 16$

167. $\frac{1}{2} x = 5 - 2x$

168. $-2 - \frac{2}{5} n = 12$

169. $\frac{5}{6} + 5n = 35$

170. $\frac{3}{10} a + 6 = 9$

5

(171.) 30m + 6.1 = 7.6   172. -4k + 16.31 = 0.31

173. 8.3y + 2.7y + 154 = 154

174. 15.8y + 0.5y - 130.4 = 16.3

1.7   LINEAR INEQUALITIES

Solve each of the following inequalities for the indicated variable.

175. 1 - 2x > 5   176. -7x ≤ 35

177. 4 + 3t > 11   178. 2y - 5 > 17

179. 55x - 250 ≤ -965   180. 5k - 3 > 2k + 9

181. -5k + 2 ≤ 2k + 9   182. x - 3 ≥ 4x - 17

183. 0.2y ≤ 0.8   (184.) -0.7k > 0.35

185. 5(x-2) ≥ 2(x+4) - 3   186. 5 - 5(2x-7) > 40

1.8   EVALUATING EXPRESSIONS

187. Determine the value of y, if y = mx + b, m = 2/3, x = 9, and b = 1/4.

188. Determine the value of y, if y = mx + b, m = 3/4, x = 8, and b = 2/3.

Evaluate the following expressions when g = 0.30, h = 1.20, and k = 0.25.

189. $g^2 + k^2$ 190. $\frac{1}{2} h^2 + \frac{1}{5} k^2$

191. $0.7h^2$ 192. h^3

(193.) $h^2k - g$ 194. 1.5gh + k

195. $k(-g^2 + 1.07h)$ 196. 0.5g(h + gk)

1.9 FRACTIONAL EQUATIONS

Solve each of the following equations for the indicated variable.

197. $\dfrac{1}{x} + \dfrac{1}{2} = 5$

198. $\dfrac{1}{y} - \dfrac{1}{2} = 4$

199. $\dfrac{5}{x} = \dfrac{6}{x} - \dfrac{1}{3}$

200. $\dfrac{5}{3x} + \dfrac{3}{x} = 1$

(201.) $\dfrac{k - 7}{k + 2} = \dfrac{1}{4}$

202. $\dfrac{b - 7}{b - 4} = \dfrac{2}{5}$

203. $\dfrac{2}{m + 1} = \dfrac{1}{m - 2}$

204. $\dfrac{5}{t - 1} = \dfrac{3}{t + 2}$

205. $3 = \dfrac{1}{4x}$

206. $5 = \dfrac{1}{2x}$

1.10 STRAIGHT LINES

(207.) Find Δx if $x_2 = 3$ and $x_1 = -1$.

208. Find Δx if $x_2 = 5$ and $x_1 = 2$.

209. Find Δy if $y_2 = 2\dfrac{1}{2}$ and $y_1 = \dfrac{3}{8}$.

210. Find Δy if $y_2 = 4\dfrac{2}{3}$ and $y_1 = 1\dfrac{1}{2}$.

211. Find Δh if $h_2 = 7.63$ and $h_1 = 5.83$.

212. Find Δh if $h_2 = 2.76$ and $h_1 = 1.44$.

Find the slope of the line passing through points P_1 and P_2.

213. $P_1 = (-2,-3)$; $P_2 = (4,9)$

(214.) $P_1 = (-1,5)$; $P_2 = (1,-3)$

215. $P_1 = (3,3)$; $P_2 = (7,3)$

216. $P_1 = (2,4)$; $P_2 = (2,-4)$

217. $P_1 = (-1,5)$; $P_2 = (2,0)$

218. $P_1 = (-1,-1)$; $P_2 = (3,4)$

Find the equation of the line determined by the given information. Express your answer in the form ax + by = c.

219. The line passing through the origin with a slope of 2.

220. The line passing through (2,4) with a slope of -3.

221. The line passing through (2,3) with a slope of 1/2.

222. The line passing through (-1,5) with a slope of 4.

223. The line passing through (4,3) with a slope of 1.

224. The line passing through (0,2) with a slope of 1/3.

Graph each of the following equations.

225. y = x + 1   226. y = -x + 2

227. x + y = 1   228. x + 2y = 6

229. 5x - 2y = 10   230. 4x - 3y = 12

231. y = 2   232. x = 3

1.11 RADICALS

Find the indicated root.

233. $\sqrt{49}$ 234. $\sqrt{81}$ 235. $\sqrt{256}$

236. $\sqrt{289}$ 237. $\sqrt{576}$ 238. $\sqrt{961}$

239. $\sqrt{484}$ 240. $\sqrt{196}$ 241. $-\sqrt{484}$

242. $-\sqrt{196}$ 243. $\sqrt{225}$ 244. $\sqrt{676}$

245. $-\sqrt{361}$ 246. $\sqrt{400}$ 247. $\sqrt[3]{64}$

248. $\sqrt[3]{-64}$ 249. $-\sqrt[3]{64}$ 250. $-\sqrt[3]{-64}$

251. $\sqrt[3]{125}$ 252. $\sqrt[3]{-125}$ 253. $-\sqrt[3]{-125}$

254. $-\sqrt[3]{125}$ 255. $\sqrt[3]{216}$ 256. $\sqrt[3]{-216}$

257. $\sqrt[4]{1,296}$ 258. $-\sqrt[4]{1,296}$

Simplify each of the following radicals.

259. $\sqrt{2}\ \sqrt{2}$ 260. $\sqrt{3}\ \sqrt{3}$ 261. $\sqrt{5}\ \sqrt{4}$

262. $\sqrt{10}\ \sqrt{2}$ 263. $\sqrt{8}\ \sqrt{2}$ 264. $\sqrt{5}\ \sqrt{10}$

265. $\sqrt{63}$ 266. $\sqrt{200}$ 267. $\sqrt{320}$

268. $\sqrt{147}$ 269. $\sqrt{243}$ 270. $\sqrt{252}$

Solve each of the following equations.

271. $x^2 = 4$ 272. $x^2 = 9$ 273. $y^2 = 256$

274. $t^2 = 900$ 275. $t^2 = 1,681$ 276. $t^2 = 1,296$

277. $x^3 = -125$ 278. $x^3 = -64$ 279. $y^3 = 125$

280. $y^3 = 216$ 281. $x^4 = 0$ 282. $x^5 = -32$

283. $t^3 = 1,000$ 284. $t^4 = 81$ 285. $x^2 - \dfrac{4}{9} = 0$

286. $y^2 - \dfrac{25}{36} = 0$ 287. $m^2 - 0.09 = 0$ 288. $x^2 - 0.49 = 0$

289. $12x^2 - \dfrac{121}{12} = 0$ 290. $64y^2 - 25 = 0$

Solve each of the following equations. Check your work.

291. $\sqrt{x} = 6$ 292. $\sqrt{x} = 4$ 293. $\sqrt{x+5} = 9$

294. $\sqrt{x} - 3 = 7$ 295. $4 + \sqrt{y-3} = 11$ 296. $6 - \sqrt{y+4} = -4$

297. $\sqrt{a/2} - 1 = 0$ 298. $\sqrt{t} - 1/2 = 2$ 299. $\sqrt{x} = 2\sqrt{5}$

300. $\sqrt{a} = 3\sqrt{4}$ 301. $\sqrt{a^2+9} = a+3$ 302. $\sqrt{b^2-16} = b - 4$

1.12 QUADRATIC EQUATIONS

Solve each of these equations by factoring.

303. $x^2 - 10x + 9 = 0$ 304. $b^2 - 7b + 10 = 0$

305. $a^2 - 2a - 35 = 0$ 306. $t^2 + t - 56 = 0$

307. $-3a^2 - 7a + 6 = 0$ 308. $-6m^2 + 23m + 4 = 0$

309. $y^2 - 9y = -18$ 310. $k^2 = 3k + 18$

311. $6b^2 + 5b = -1$ 312. $6x^2 + 15 = 19x$

Using the quadratic formula, solve each of these equations.

313. $x^2 + 2x - 3 = 0$ 314. $y^2 + 4y - 5 = 0$

315. $a^2 + 8a - 9 = 0$ 316. $m^2 - 2m - 8 = 0$

317. $w^2 = 4w - 4$ 318. $d^2 = 8d - 16$

319. $x^2 - 5x - 6 = 0$ 320. $y^2 + 7y - 8 = 0$

321. $x^2 - 2x - 5 = 0$ 322. $y^2 + 6y - 5 = 0$

Sketch the graph of each of the following equations.

323. $y = 2x^2$ 324. $y = 3x^2$

325. $y = -3x^2$ 326. $y = -4x^2$

327. $y = -x^2 + 3$ 328. $y = -x^2 - 2$

329. $y = x^2 + 1$ 330. $y = x^2 + 5$

331. $y = x^2 + 2x + 1$ 332. $y = x^2 + 4x + 4$

333. $y = -x^2 + 6x - 9$ 334. $y = x^2 - 2x + 1$

1.13 EXPONENTIAL EQUATIONS

Graph each of the following equations.

335. $y = 3^x$ 336. $y = 4^x$

337. $y = (\frac{1}{3})^x$ 338. $y = (\frac{1}{4})^x$

339. $y = (\frac{1}{3})^{-x}$ 340. $y = (\frac{1}{4})^{-x}$

341. $y = (\frac{2}{3})^{-x}$ 342. $y = (\frac{3}{4})^{-x}$

343. $y = e^{2x}$ 344. $y = e^{1/2x}$

Solve the following exponential equations.

345. $2^{x+3} = 16$ 346. $5^{x-10} = 125$

347. $3^{2y-3} = 243$ 348. $4^{x-5} = 4^{-1}$

349. $2^{5x+2} = 2^{3x-4}$ 350. $6^{x-2} = 6^{3x-4}$

351. $3^x = (\frac{1}{3})^{x-6}$ 352. $(\frac{1}{8})^x = 2^{x-8}$

353. $2^{3x+9} = 8^{-x+1}$ 354. $3^{3x} = 9^{x+5}$

355. $81^{x+1} = 27^{2x-1}$ 356. $125^{2x-1} = 625^x$

357. $2^{x^2} = 32$ 358. $3^{x^2} = 243$

359. $2^{x^2-3} = 64$ 360. $7^{x^2-4} = 49$

1.14 LOGARITHMS

Express each logarithmic statement in exponential form.

361. $\log_2 16 = 4$ 362. $\log 1{,}000 = 3$

363. $\log_3 81 = 4$ 364. $\log_7 343 = 3$

365. $\log_3 243 = 5$ 366. $\log_5 625 = 4$

367. $\log 0.1 = -1$ 368. $\log_5 0.04 = -2$

369. $\log_7 1 = 0$ 370. $\log_9 81 = 2$

371. $\log_8 \frac{1}{64} = -2$ 372. $\log_3 \frac{1}{27} = -3$

373. $\log_9 3 = \frac{1}{2}$ 374. $\log_{16} 4 = \frac{1}{2}$

375. $\log_8 4 = \frac{2}{3}$ 376. $\log_9 27 = \frac{3}{2}$

Express each exponential statement in logarithmic form.

377. $10^4 = 10{,}000$ 378. $4^3 = 64$ 379. $2^7 = 128$

380. $5^3 = 125$ 381. $5^0 = 1$ 382. $2^5 = 32$

11

383. $3^4 = 81$ 384. $2^8 = 256$ 385. $4^{1/2} = 2$

386. $25^{1/2} = 5$ 387. $8^{2/3} = 4$ 388. $16^{3/4} = 8$

389. $3^{-3} = \dfrac{1}{27}$ 390. $4^{-2} = \dfrac{1}{16}$

391. $10^{-5} = 0.00001$ 392. $10^{-6} = 0.000001$

Find the indicated logarithm.

393. $\log_9 1$ 394. $\log_7 7$

395. $\log_{17} 289$ 396. $\log_2 256$

397. $\log 0.001$ 398. $\log 100,000$

399. $\log_8 512$ 400. $\log_{11} 1331$

401. $\log_9 \dfrac{1}{81}$ 402. $\log_5 \dfrac{1}{625}$

403. $\log_{64} 16$ 404. $\log_{32} 8$

405. $\log_2 128$ 406. $\log_3 \dfrac{1}{27}$

407. $\log 0.0001$ 408. $\log \dfrac{1}{100}$

Solve the equation for x.

409. $\log_x 64 = 3$ 410. $\log_x 81 = 2$

411. $\log_5 625 = x$ 412. $\log_4 256 = x$

413. $\log_8 1 = x$ 414. $\log_3 x = 3$

415. $\log_3 x = -2$ 416. $\log_2 x = -3$

417. $\log_{27} x = \dfrac{1}{3}$ 418. $\log_2 x = -2$

419. $\log_x 81 = 4$ 420. $\log_x 6 = \dfrac{1}{2}$

421. $\log_4 x = -2$ 422. $\log_x 81 = -2$

423. $\log 0.001 = x$ 424. $\log 0.0001 = x$

425. Graph the equation $y = \log_3 x$.

426. Graph the equation $y = \log_4 x$.

427. Graph the equation $y = \log x$.

428. Graph the equation $y = \log_8 x$.

If the $\log 2 = 0.3010$, $\log 3 = 0.4771$, and $\log 5 = 0.6990$, find each of the following logs.

429. $\log 18$

430. $\log 16$

431. $\log 27$

432. $\log 20$

433. $\log 50$

434. $\log 32$

435. $\log 125$

436. $\log 1.5$

437. $\log 2.5$

438. $\log 0.9$

439. $\log 0.4$

440. $\log 0.27$

441. $\log 0.16$

442. $\log \frac{10}{3}$

443. $\log \frac{3}{5}$

444. $\log \frac{4}{15}$

Solve the logarithmic equation. Check your solution.

445. $\log x - \log 3 = 0$

446. $\log (x + 1) - \log 7 = 0$

447. $\log (x - 2) - \log 5 = 1$

448. $\log 4 + \log x = 1$

449. $\log_3(x - 4) + \log_3(x + 4) = 2$

450. $\log_7(x - 5) + \log_7(x + 1) = 1$

451. $\log(x - 1) + \log(x + 2) = \log 4$

452. $\log(x - 4) + \log(x + 1) = \log 6$

453. $\log(x - 7) = \log(3x - 5) + \log 3$

454. $\log_6 10 + \log_6(2x - 7) = \log_6(3x - 19)$

455. $\log_5(6x + 4) = \log_5(6x - 1) + \log_5 2$

456. $\log_8(4x - 5) + \log_8 3 = \log_8(2x + 5)$

457. $\log_3(x + 4) - \log_3(x - 1) = \log_3 x$

458. $\log(3x + 1) - \log x = \log(x + 1)$

459. $\log(3x + 1) - \log(3 + x) = \log 2$

460. $\log(4x + 5) - \log(x + 3) = \log 3$

Stop here if you are enrolling in a calculus course that is not designed for mathematics and physics majors. If you plan to major in mathematics or physics, continue to the end.

1.15 TRIGONOMETRY

461. What is the reciprocal of sec θ?

462. What is the reciprocal of csc θ?

463. If sin $\theta = \dfrac{1}{2}$, what is the value of $\sin^5\theta$?

464. If tan $\theta = \dfrac{3}{2}$, what is the value of $\tan^4\theta$?

465. What trigonometric ratio is equal to $\dfrac{\sin \theta}{\cos \theta}$?

466. What trigonometric ratio is equal to $\dfrac{\cos \theta}{\sin \theta}$?

467. Complete: $\sin^2\theta + \cos^2\theta = ?$

468. Complete: $1 + \tan^2\theta = ?$

469. Evaluate $\sin^4 45°$

470. Evaluate $\tan^6 45°$

471. Evaluate $\sin^3 \dfrac{\pi}{6}$

472. Evaluate $\sin^2 \dfrac{\pi}{3}$

473. Evaluate $\tan^3 \dfrac{\pi}{4}$

474. Evaluate $\cos^4 \dfrac{\pi}{3}$

475. Two sides of a triangle are 12 inches long and they include an angle of 120°. Find the length of the third side of the triangle.

476. Two sides of a triangle are each 10 inches long and they include an angle of 60°. Find the length of the third side of the triangle.

477. Sin 90° = ?

478. Cos 180° = ?

479. Cos 135° = ?

480. Sin 150° = ?

481. Sec 135° = ?

482. Csc 150° = ?

483. Cos 240° = ?

484. Sin 300° = ?

485. $\cos(-\frac{\pi}{4})$ = ?

486. $\sin(-\frac{\pi}{6})$ = ?

487. $\cos \frac{3\pi}{2}$ = ?

488. $\sin \frac{3\pi}{2}$ = ?

489. Graph y = sin 2x

490. Graph y = cos 2x

491. Graph y = 3cos x

492. Graph y = 4sin x

For each of the following equations state the amplitude and the period.

493. y = sin x

494. y = cos x

495. y = cos 4x

496. y = sin 3x

497. $y = \sin \frac{x}{3}$

498. $y = \cos \frac{x}{4}$

499. y = 3sin(2x)

500. y = 4cos(3x)

1.16 FUNCTIONS

501. If $f(x) = -3x^2 + x - 5$, determine $f(-1)$.

502. If $g(x) = \sqrt{2x + 9}$, determine $g(8)$.

503. If $f(x) = x^3 + 1$ and $g(x) = \sqrt{x}$, determine $f[g(4)]$.

504. If $f(x) = x^3 + 1$ and $g(x) = \sqrt{x}$, determine $g[f(2)]$.

505. If $g(x) = [x]$, where the brackets denote the greatest integer function, what is the value of $g(5.71)$?

506. If $g(x) = [x]$, where the brackets denote the greatest integer function, what is the value of $g(-2.05)$?

15

507. What is the domain of $h(x) = \sqrt{2x - 1}$?

508. What is the domain of $h(x) = \dfrac{x + 5}{2x - 1}$?

509. What is the domain of $f(x) = \dfrac{x}{x^2 - 4}$?

510. What is the domain of the exponential function?

511. What is the domain of the natural logarithm function?

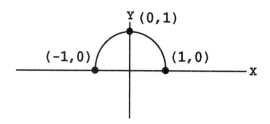

512. For the function f(x) graphed above, what is the domain?

513. For the function f(x) graphed above, what is the range?

514. What is the range of $g(x) = \sqrt{x} + 2$?

515. What is the range of $g(x) = x^{1/2} - 1$?

516. What is the range of the natural logarithm function?

517. What is the range of the exponential function?

518. What is the range of $h(x) = -x^2 + 4$?

519. What is the range of $h(x) = 3x^2 + 5$?

520. What is the range of the absolute value function?

1.17 **PROBLEM SOLVING**

521. How many prime numbers are there between 0 and 10 inclusive?

522. How many composite numbers are there between 0 and 10 inclusive?

523. A recipe for 50 people uses 3 cups of sugar; how many cups of sugar are needed for 75 people?

524. A dieter lost 2 pounds in three weeks. If he continues to lose at this rate, how many weeks will it take him to lose 24 pounds?

525. A man has taken out a simple interest loan on a principal of $28,000 for 2.50 years. If he pays interest of $9,800, what is his rate of interest?

526. After a period of 3 years, a woman's investment at 12.5% has amounted to a total of $68,750. What was the principal that she invested 3 years ago?

527. If one side of a triangle has a length of 5 inches and a second side has a length of 8 inches, what length is possible for the third side of this triangle?

528. If one side of a triangle has a length of 6 inches and a second side has a length of 10 inches, what length is possible for the third side of this triangle?

529. There are five people in a room and each person wishes to shake hands with every other person exactly one time. How many handshakes are required?

530. In how many ways can one place three books "a", "b", and "c" on a shelf so that reading, left to right, the books will not be in alphabetical order?

531. Determine three consecutive odd-numbers whose sum is 81.

532. Determine three consecutive even-numbers whose sum is 102.

533. Your combined Federal and state taxes amounts to 35% of your gross pay. What is the minimum gross pay, in dollars, that you must earn in order to have a net pay of at least $40,000?

534. A class of 31 students was organized for a tennis tournament. After each girl in the class was assigned to a boy as a partner, there were 7 boys left. How many girls are in the class?

535. Each side of an equilateral triangle is 4 inches longer than the side of a square. The sum of the perimeters of the two figures is 96 inches. How long is each side of the triangle?

536. How many degrees are in an angle that is 1/4 as large as its supplement?

537. Two exterior angles of a triangle total 200°; how many degrees are in the third exterior angle?

538. If two adjacent angles whose sum is 96° are bisected, how many degrees are in the angle formed by the two bisectors?

539. How many degrees are there in the angle formed by the bisectors of the acute angles of a 30°, 60°, right triangle?

540. The base of an isosceles triangle is the same length as the side of a square; each of the equal sides of the triangle is 3 inches longer than the length of a side of this square. If the sum of the perimeters of the two figures is 90 inches, how long is each side of the triangle?

541. John is three years older than Nancy, and Nancy is twice as old as Norman. The average of their ages is 16. How old is each person?

542. One number is 10 greater than another and their average is less than or equal to 24. What is the greatest value that the smaller of the two numbers can have?

543. Steve has more than $2.00 in nickels and dimes. He has 10 more dimes than nickels. At least how many dimes does he have?

544. Lisa has two fewer nickels than dimes and two more quarters than dimes. If the total of her change is $2.80, how many of each type of coin does she have?

545. At noon a plane leaves an airport and flies due west at 540 mi/hr. At 1 P.M. a second plane leaves the same airport and also flies due west at 720 mi/hr. At what time does the second plane overtake the first plane?

546. Paula starts walking from a specific location at the rate of 4 mi/hr. Five hours later Patricia starts walking from the same location at the rate of 6 mi/hr. How many hours will Patricia walk before she overtakes Paula?

547. What was the principal on a simple interest loan written at a rate of 12.5% for two years if the person received a total of $22,500 at the end of the two years?

548. What was the principal on a simple interest loan written at a rate of 16% for 2.25 years if the person received a total of $68,000 at the end of 2.25 years?

549. Grant and Bloom are partners who divide their profits in the ratio 3:5. Find Grant's profit if Bloom receives $15,000.

550. A store has kept the ratio of management level people to sales people at 2:9. If the store has 180 sales people, how many management level people are there?

551. A coffee merchant wants to make a blend of 12 pounds of coffee to sell at $6 per pound. The blend is to be made by combining an expensive $10 per pound coffee with a modestly priced $4 per pound coffee. How many pounds of each type of coffee are needed?

552. A solution of 60 quarts of sugar and water is 20% sugar. How much water must be added to make a solution that is 5% sugar?

553. If the radius of a circle is decreased by 6 inches, the area of the circle is divided by 9. Find the original radius.

554. A deck 3 meters wide surrounds a rectangular pool which is twice as long as it is wide. Find the dimensions of the pool if the area of the deck is 360 square meters.

555. One candle will burn up completely in 4 hours while a second candle of equal length requires 5 hours to burn up completely. If the candles are lit at the same time, how long will they burn before one of them is exactly three times as long as the other?

556. Two key punch operators work at different rates. Working together they can process a payroll in 15 hours. After one operator has worked 5 hours at the job, the second operator joins her. Together they finish the job in 13 hours. How long would it have taken the first operator working alone to have finished the payroll?

557. A student has not studied and must take a 10 question multiple-choice exam in which each question has 5 possible answers. In how many ways can she mark her exam?

558. How many automobile license plates can be made if each plate begins with two letters of the alphabet and ends with three digits between 0 and 9 inclusive? It is assumed that repetition of letters or digits is allowed.

559. A river steamer cruises 58 kilometers downstream in the same time that it cruises 40 kilometers upstream. The engine drives the steamer in calm water at a rate of 20 km/hr more than the rate of the current. Find the rate of the current.

560. In calm water, a motorboat can travel 4 times as fast as the current in a large river. A trip up the river and back totaling 90 kilometers required 8 hours. Find the rate of the current.

1. $\frac{5}{6}$ 2. 1 3. $1\frac{3}{8}$ 4. $\frac{13}{36}$

5. $\frac{11}{14}$ 6. $1\frac{3}{16}$ 7. $\frac{73}{80}$ 8. $\frac{17}{28}$

9. $\frac{29}{36}$ 10. $\frac{83}{112}$ 11. $\frac{2}{7}$ 12. $\frac{6}{25}$

13. $\frac{1}{4}$ 14. $\frac{3}{4}$ 15. $\frac{7}{24}$ 16. $\frac{23}{42}$

17. $9\frac{1}{8}$ 18. $\frac{19}{56}$ 19. $3\frac{9}{10}$ 20. $\frac{20}{63}$

21. $\frac{6}{11}$ 22. $\frac{1}{8}$ 23. $\frac{5}{8}$ 24. $6\frac{2}{3}$

25. $\frac{16}{27}$ 26. $5\frac{1}{10}$ 27. 1 28. $\frac{2}{5}$

29. 6 30. $1\frac{1}{2}$ 31. 1 32. 5

33. 9 34. $\frac{1}{4}$ 35. 4 36. $\frac{5}{16}$

37. $\frac{1}{100}$ 38. $1\frac{3}{5}$ 39. $\frac{7}{16}$ 40. $1\frac{2}{3}$

41. 1.2 42. 11 43. 18.5 44. 20.4

45. 58.39 46. 276.94 47. 2.9 48. 670.38

49. 0.063 50. 71.204 51. 0.6 52. 2.5

53. 1.7 54. 6.6 55. 542.38 56. 104.88

57. 80.93 58. 54.49 59. 18.57 60. 54.25

61. 24 62. 41.6 63. 0.84 64. 22.09

65. 5.08 66. 323.96 67. 10.8 68. 19,500

69. 4.444 70. 77.004 71. 4.3 72. 1.98

73. 7.2 74. 4.8 75. 4.14 76. 4.71

77. 0.22 78. 4.2 79. 8.2 80. 6.1

81. 1:2 82. 1:3 83. 9:4 84. 4:5

85. 1:4 86. 7:17 87. 4 and 15 88. 2 and 21

89. 35 90. 14 91. 2 92. 15

93. 25 94. 36 95. 63 96. 25

97. 40 inches 98. 22 pounds 99. 98 100. 80

101. 1 102. $\frac{1}{8}$ 103. $\frac{1}{125}$ 104. $\frac{1}{32}$

105. 128 106. 0.064 107. $\frac{9}{16}$ 108. $15\frac{5}{8}$

109. $\frac{1}{9}$ 110. $\frac{8}{27}$ 111. 3^6 112. 10^5

113. $(-2)^8$ 114. x^{12} 115. 2^{20} 116. 3^9

117. $\frac{1}{10^2}$ 118. 4^4 119. $\frac{1}{8^4}$ 120. 4^5

121. 6^4 122. x^3 123. $\frac{1}{2^2}$ 124. $\frac{1}{8^4}$

125. $3x(x + 2y + 3x^2)$ 126. $a(x^2 - 3x + 5a)$

127. $(5y - 3)(5y + 3)$ 128. $(10x - 1)(10x + 1)$

129. prime 130. prime

131. $2(3x - 7y)(3x + 7y)$ 132. $7(x - 2y)(x + 2y)$

133. $(x + y)(x^2 - xy + y^2)$ 134. $(w + 1)(w^2 - w + 1)$

135. $(t - 3)(t^2 + 3t + 9)$ 136. $(a - b)(a + b)(a^2 + b^2)$

137. prime 138. prime

139. $(x - 5)(x - 3)$ 140. $(b + 3)(b + 7)$

141. $(7k + 4)(5k - 1)$ 142. $(3p + 5)(2p + 1)$

143. $(2k + 7)(2k - 3)$ 144. prime

145. prime 146. $(2t - 1)(t - 1)$

147. $(8x - 5)(5x + 8)$ 148. $(8y - 7)(7y + 8)$

149. $(3x + 2)(4x - 7)$ 150. $(4a + 7)^2$

151. $(4x - 7)(3x - 2)$ 152. $(5r - 2)(2r + 1)$

153. $2(y - 1)(2y - 3)$

154. $3(2x + 1)(x + 3)$

155. $5(a - b)(a + b)(a^2 + b^2)$

156. $7(x - y)(x + y)(x^2 + y^2)$

157. $x = 1$

158. $x = 1$

159. $x = 2\frac{2}{3}$

160. $x = 5$

161. $x = -8$

162. $x = 10$

163. $n = -\frac{5}{6}$

164. $b = 5$

165. $x = 62$

166. $b = 28$

167. $x = 2$

168. $n = -35$

169. $n = 6\frac{5}{6}$

170. $a = 10$

171. $m = 0.05$

172. $k = 4$

173. $y = 0$

174. $y = 9$

175. $x < -2$

176. $x \geq -5$

177. $t > 2\frac{1}{3}$

178. $y > 11$

179. $x \leq -13$

180. $k > 4$

181. $k \geq -1$

182. $x \leq 4\frac{2}{3}$

183. $y \leq 4$

184. $k < -0.5$

185. $x \geq 5$

186. $x < 0$

187. $6\frac{1}{4}$

188. $6\frac{2}{3}$

189. 0.1525

190. 0.7325

191. 1.008

192. 1.728

193. 0.06

194. 0.79

195. 0.2985

196. 0.19125

197. $\frac{2}{9}$

198. $\frac{2}{9}$

199. 3

200. $4\frac{2}{3}$

201. 10

202. 9

203. 5

204. $-6\frac{1}{2}$

205. $\frac{1}{12}$

206. $\frac{1}{10}$

207. 4

208. 3

209. $2\frac{1}{8}$

210. $3\frac{1}{6}$

211. 1.80

212. 1.32

213. 2

214. -4

215. 0

216. undefined

217. $-\frac{5}{3}$

218. $\frac{5}{4}$

219. $2x-y = 0$

220. $3x+y = 10$

221. $x-2y = -4$

222. $4x-y = -9$

223. $x-y = 1$

224. $x-3y = -6$

225.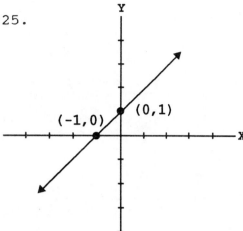

226.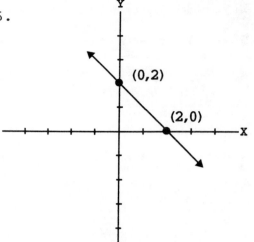

227.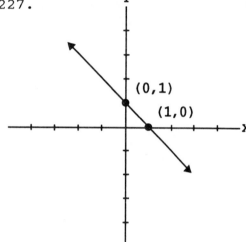

228.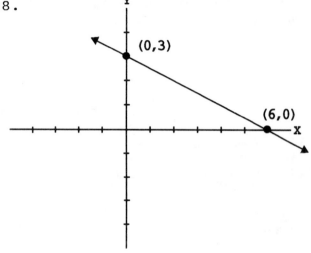

229.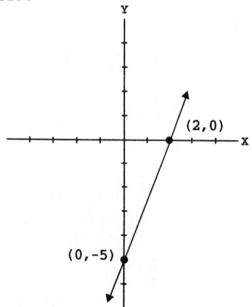

230.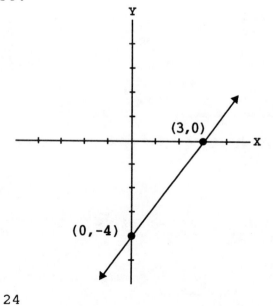

24

231.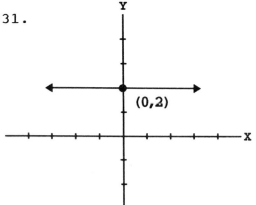

(0,2)

232.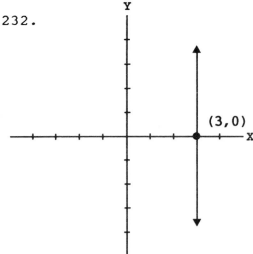

(3,0)

233. 7

234. 9

235. 16

236. 17

237. 24

238. 31

239. 22

240. 14

241. -22

242. -14

243. 15

244. 26

245. -19

246. 20

247. 4

248. -4

249. -4

250. 4

251. 5

252. -5

253. 5

254. -5

255. 6

256. -6

257. 6

258. -6

259. 2

260. 3

261. $2\sqrt{5}$

262. $2\sqrt{5}$

263. 4

264. $5\sqrt{2}$

265. $3\sqrt{7}$

266. $10\sqrt{2}$

267. $8\sqrt{5}$

268. $7\sqrt{3}$

269. $9\sqrt{3}$

270. $6\sqrt{7}$

271. ± 2

272. ± 3

273. ± 16

274. ± 30

275. ± 41

276. ± 36

277. -5

278. -4

279. 5

280. 6

281. 0

282. -2

283. 10

284. ± 3

285. $\pm \dfrac{2}{3}$

286. $\pm \dfrac{5}{6}$

287. ± 0.3

288. ± 0.7

289. $\pm \dfrac{11}{12}$

290. $\pm \dfrac{5}{8}$

291. 36

292. 16

293. 76

294. 100

295. 52

296. 96

297. 2

298. $\frac{25}{4}$

299. 20

300. 36

301. 0

302. 4

303. 1,9

304. 2,5

305. -5,7

306. -8,7

307. -3, $\frac{2}{3}$

308. $-\frac{1}{6}$,4

309. 6,3

310. -3,6

311. $-\frac{1}{2}$, $-\frac{1}{3}$

312. $1\frac{1}{2}$, $1\frac{2}{3}$

313. -3,1

314. -5,1

315. -9,1

316. -2,4

317. 2(double root)

318. 4(double root)

319. -1,6

320. -8,1

321. $1 \pm \sqrt{6}$

322. $-3 \pm \sqrt{14}$

323.

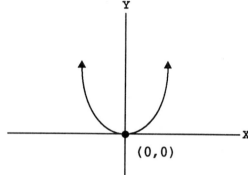

324.

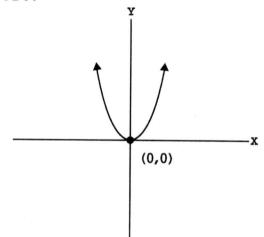

325.

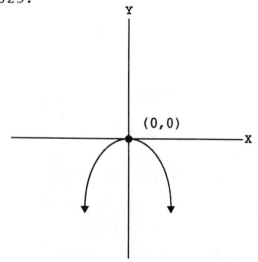

326.

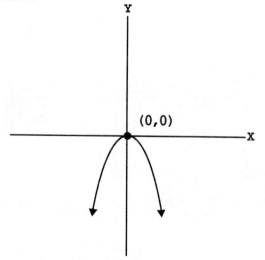

327.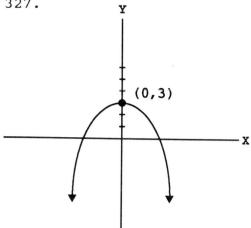

328.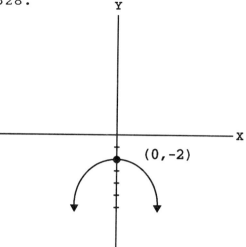

329.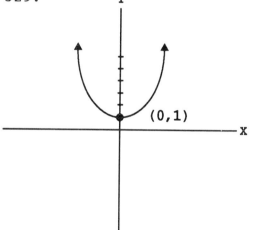

330.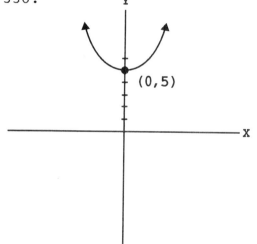

331.

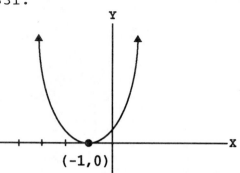

(-1,0)

332.

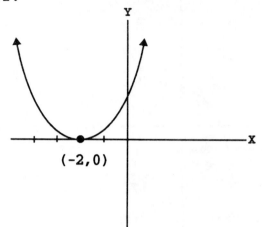

(-2,0)

333.

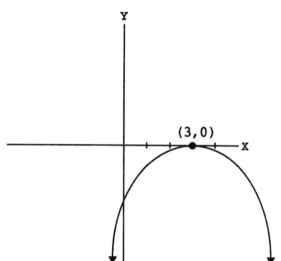

(3,0)

334.

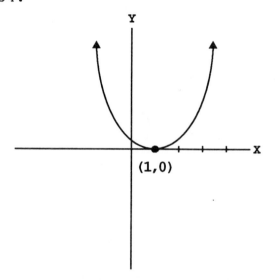

(1,0)

335.

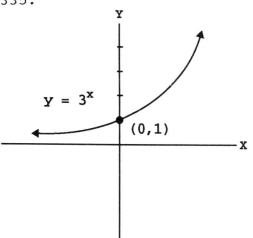

$Y = 3^x$

(0,1)

336.

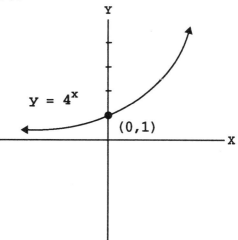

$Y = 4^x$

(0,1)

337.

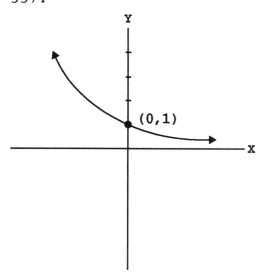

(0,1)

338.

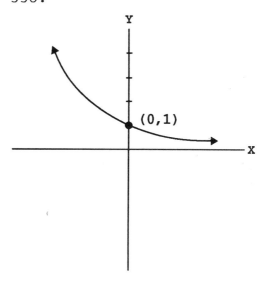

(0,1)

339.

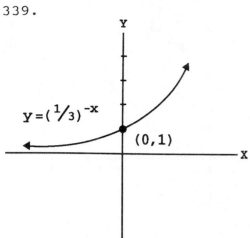

$y = (1/3)^{-x}$

(0,1)

340.

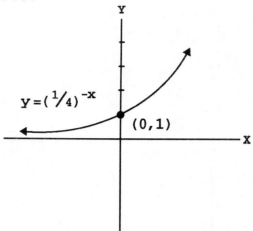

$y = (1/4)^{-x}$

(0,1)

341.

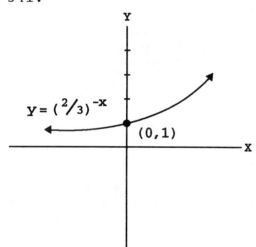

$y = (2/3)^{-x}$

(0,1)

342.

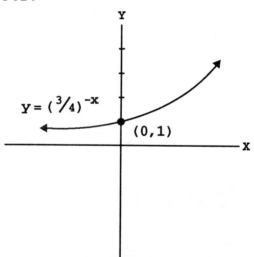

$y = (3/4)^{-x}$

(0,1)

30

343.

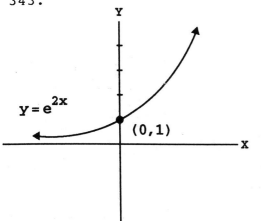

344.

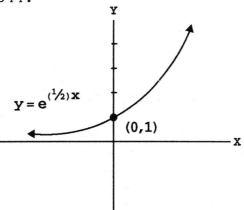

345. $x = 1$

346. $x = 13$

347. $y = 4$

348. $x = 4$

349. $x = -3$

350. $x = 1$

351. $x = 3$

352. $x = 2$

353. $x = -1$

354. $x = 10$

355. $x = \frac{7}{2}$

356. $x = \frac{3}{2}$

357. $x = \pm \sqrt{5}$

358. $x = \pm \sqrt{5}$

359. $x = \pm 3$

360. $x = \pm \sqrt{6}$

361. $16 = 2^4$

362. $1,000 = 10^3$

363. $81 = 3^4$

364. $343 = 7^3$

365. $243 = 3^5$

366. $625 = 5^4$

367. $0.1 = 10^{-1}$

368. $0.04 = 5^{-2}$

369. $1 = 7^0$

370. $81 = 9^2$

371. $\frac{1}{64} = 8^{-2}$

372. $\frac{1}{27} = 3^{-3}$

373. $3 = 9^{1/2}$

374. $4 = 16^{1/2}$

375. $4 = 8^{2/3}$

376. $27 = 9^{3/2}$

377. $\log 10,000 = 4$

378. $\log_4 64 = 3$

379. $\log_2 128 = 7$

380. $\log_5 125 = 3$

381. $\log_5 1 = 0$

382. $\log_2 32 = 5$

383. $\log_3 81 = 4$

384. $\log_2 256 = 8$

385. $\log_4 2 = \frac{1}{2}$

386. $\log_{25} 5 = \frac{1}{2}$

387. $\log_8 4 = \frac{2}{3}$

388. $\log_{16} 8 = \frac{3}{4}$

389. $\log_3 \frac{1}{27} = -3$

390. $\log_4 \frac{1}{16} = -2$

391. $\log 0.00001 = -5$

392. $\log 0.000001 = -6$

393. 0

394. 1

395. 2

396. 8

397. -3

398. 5

399. 3

400. 3

401. -2

402. -4

403. $\frac{2}{3}$

404. $\frac{3}{5}$

405. 7

406. -3

407. -4

408. -2

409. 4

410. 9

411. 4

412. 4

413. 0

414. 27

415. $\frac{1}{9}$

416. $\frac{1}{8}$

417. 3

418. $\frac{1}{4}$

419. 3

420. 36

421. $\frac{1}{16}$ 422. $\frac{1}{9}$ 423. -3 424. -4

425.

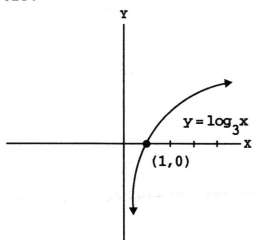

$y = \log_3 x$

$(1,0)$

426.

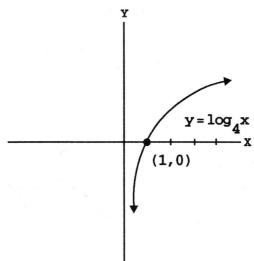

$y = \log_4 x$

$(1,0)$

427.

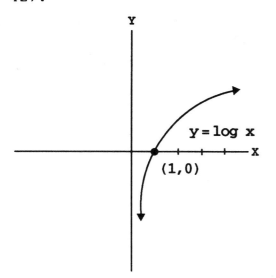

$y = \log x$

$(1,0)$

428.

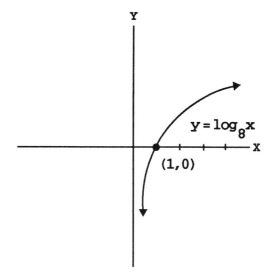

$y = \log_8 x$

$(1,0)$

33

429. 1.2552	430. 1.2040	431. 1.4313
432. 1.301	433. 1.6990	434. 1.5050
435. 2.097	436. 0.1761	437. 0.398
438. -0.0458	439. -0.398	440. -0.5687
441. -0.796	442. 0.5229	443. -0.2219
444. -0.5741	445. x = 3	446. x = 6
447. x = 52	448. x = $\frac{5}{2}$	449. x = 5
450. x = 6	451. x = 2	452. x = 5
453. no solution	454. no solution	455. x = 1
456. x = 2	457. x = $1 + \sqrt{5}$	458. x = $1 + \sqrt{2}$
459. x = 5	460. x = 4	461. cos θ
462. sin θ	463. $\frac{1}{32}$	464. $\frac{81}{16}$
465. tan θ	466. ctn θ	467. 1
468. $\sec^2\theta$	469. $\frac{1}{4}$	470. 1
471. $\frac{1}{8}$	472. $\frac{3}{4}$	473. 1
474. $\frac{1}{16}$	475. $12\sqrt{3}$	476. 10
477. 1	478. -1	479. $-\frac{\sqrt{2}}{2}$
480. $\frac{1}{2}$	481. $-\sqrt{2}$	482. 2
483. $-\frac{1}{2}$	484. $-\frac{\sqrt{3}}{2}$	485. $\frac{\sqrt{2}}{2}$
486. $-\frac{1}{2}$	487. 0	488. -1

489.

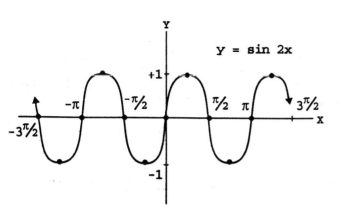

490.

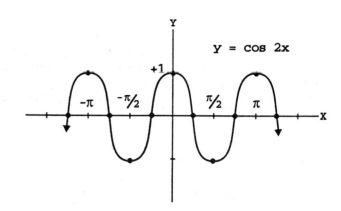

491.

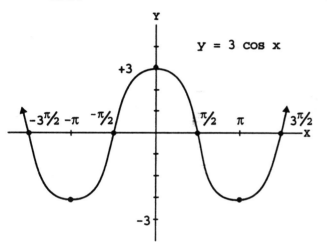

492.

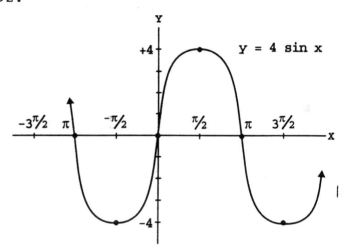

493. 1; 2π

494. 1; 2π

495. 1; $\dfrac{\pi}{2}$

496. 1; $\dfrac{2\pi}{3}$

497. 1; 6π

498. 1; 8π

499. 3; π

500. 4; $\dfrac{2\pi}{3}$

501. -9

502. 5

503. 9

504. 3

505. 5

506. -3

507. $x \geq \dfrac{1}{2}$

508. all reals except 1/2   509. all reals except +2 and -2

510. all reals   511. x > 0   512. -1 ≤ x ≤ 1

513. 0 ≤ y ≤ 1   514. y ≥ 2   515. y ≥ -1

516. all reals   517. y > o   518. y ≤ 4

519. y ≥ 5   520. y ≥ 0   521. there are 4 primes

522. there are 5 composites   523. 4 1/2 cups

524. 36 weeks   525. 14%

526. $50,000

527. The length of the third side is greater than 3 inches and less than 13 inches.

528. The length of the third side is greater than 4 inches and less than 16 inches.

529. 10   530. 5

531. 25, 27, and 29   532. 32, 34, and 36

533. $61,539   534. 12 girls

535. 16 inches   536. 36°

537. 160°   538. 48°   539. 135°

540. 12 inches, 15 inches, and 15 inches

541. Norman is 9, Nancy is 18, and John is 21.

542. less than or equal to 19 543. 17

544. 4 nickels, 6 dimes, and 8 quarters

545. 4 P.M.   546. 10 hours

547. $18,000   548. $50,000

549. $9,000   550. 40

551. 4 pounds of the $10 coffee and 8 pounds of the $4 coffee

552. 180 quarts   553. 9 inches

554. 18 meters and 36 meters 555. 3 7/11 hours

556. 37 $\frac{1}{2}$ hours   557. 5^{10}

558. 676,000   559. 4.5 km/hr

560. 3 km/hr

PART II

A REVIEW OF BASIC CONCEPTS

2.1 FRACTIONS

A **fraction** is a **whole number** (0,1,2,3,...) divided by a **natural number** (1,2,3,...). The number above the fraction line is called the **numerator** and the number below the line is called the **denominator**.

$$\frac{3}{7} \begin{array}{l} \leftarrow \text{numerator} \\ \leftarrow \text{denominator} \end{array}$$

A fraction is proper when the numerator is less than (<) the denominator. Some examples of **proper fractions** are:

$$\frac{3}{8} \, , \, \frac{5}{7} \, , \, \frac{1}{9} \, , \, \frac{11}{12} \, , \text{ and } \frac{101}{140} \, .$$

A fraction is improper when the numerator is greater than or equal to (≥) the denominator. Some examples of **improper fractions** are:

$$\frac{3}{1} \, , \, \frac{8}{7} \, , \, \frac{5}{2} \, , \, \frac{9}{4} \, , \, \frac{8}{8} \, , \text{ and } \frac{40}{30} \, .$$

How to compare fractional values:

1. $\frac{a}{b} = \frac{c}{d}$ whenever $a \cdot d = b \cdot c$

2. If $a \cdot d < b \cdot c$, then $\frac{a}{b} < \frac{c}{d}$

3. If $a \cdot d > b \cdot c$, then $\frac{a}{b} > \frac{c}{d}$

Example 1. Are the fractions $\frac{3}{5}$ and $\frac{6}{10}$ equal?

Solution: a = 3, b = 5, c = 6, and d = 10

$$3 \cdot 10 \; ? \; 5 \cdot 6$$

$$30 = 30$$

Thus $\frac{3}{5} = \frac{6}{10}$

Example 2. Which fraction is larger $\frac{9}{10}$ or $\frac{4}{5}$?

Solution: a = 9, b = 10, c = 4, and d = 5.

$\qquad$ 9 · 5 ? 10 · 4

$\qquad$ 45 > 40

$\qquad$ Therefore $\frac{9}{10}$ > $\frac{4}{5}$

A **mixed number** is formed when a whole number is added to a proper fraction. 4 1/3 is a mixed number; it means 4 + 1/3. Similarly 2 4/5 is shorthand for 2 + 4/5.

To convert a mixed number to an improper fraction, multiply the whole number by the denominator of the fraction and add the numerator of the fraction. This will give you your new numerator; your denominator is the denominator of the fractional part of the mixed number.

Example 3. Convert $2\frac{3}{4}$ to an improper fraction.

Solution: $2\frac{3}{4} = \frac{(2 \times 4) + 3}{4} = \frac{11}{4}$

To convert an improper fraction into a mixed number, divide the denominator into the numerator. This produces the whole number portion of the answer. The remainder is the numerator of the fractional portion of your answer. The denominator of the fractional portion is the original denominator.

Example 4. Convert $\frac{13}{5}$ to a mixed number.

Solution:
$$5)\overline{13}$$
$$\frac{10}{3}$$
with quotient 2

$\qquad$ So $\frac{13}{5} = 2\frac{3}{5}$

To **add** (or **subtract**) fractions with the same denominator, add (or subtract) the numerators and place the answer over the common denominator.

Example 5.   a.   Add: $\dfrac{4}{11} + \dfrac{5}{11}$

               b.   Subtract: $\dfrac{7}{9} - \dfrac{5}{9}$

Solution: a. $\dfrac{4}{11} + \dfrac{5}{11} = \dfrac{4 + 5}{11} = \dfrac{9}{11}$

            b. $\dfrac{7}{9} - \dfrac{5}{9} = \dfrac{7 - 5}{9} = \dfrac{2}{9}$

To **add** (or **subtract**) fractions with unlike denominators, begin by finding the least common denominator (LCD) of the given denominators. Convert the given fractions to equivalent fractions with the LCD as the denominator. Add (or subtract) the fractions and reduce the answer to lowest terms.

Example 6. Add: $\dfrac{3}{7} + \dfrac{2}{21}$

Solution: $\dfrac{3}{7} + \dfrac{2}{21} = \dfrac{9}{21} + \dfrac{2}{21} = \dfrac{9 + 2}{21} = \dfrac{11}{21}$

Example 7. Subtract: $\dfrac{3}{5} - \dfrac{1}{4}$

Solution: $\dfrac{3}{5} - \dfrac{1}{4} = \dfrac{12}{20} - \dfrac{5}{20} = \dfrac{12 - 5}{20} = \dfrac{7}{20}$

Example 8. Add: $\dfrac{1}{6} + \dfrac{7}{10} + \dfrac{2}{15}$

Solution: $\dfrac{1}{6} + \dfrac{7}{10} + \dfrac{2}{15} = \dfrac{5}{30} + \dfrac{21}{30} + \dfrac{4}{30}$

$$= \dfrac{5 + 21 + 4}{30}$$

$$= \dfrac{30}{30} \text{ or } 1$$

Example 9. Add: $26 \frac{4}{15}$

$$+ 42 \frac{1}{3}$$

Solution: $26 \frac{4}{15}$ $=$ $26 \frac{4}{15}$

$+ 42 \frac{1}{3}$ $=$ $+ 42 \frac{5}{15}$

$68 \frac{9}{15} = 68 \frac{3}{5}$

EXAMPLE 10.   Subtract: $6 \frac{1}{6}$

$$- 3 \frac{7}{8}$$

Solution: $6 \frac{1}{6} = 6 \frac{4}{24} = 5 \frac{28}{24}$; here we "borrowed"

$- 3 \frac{7}{8} = 3 \frac{21}{24} = 3 \frac{21}{24}$

$$2 \frac{7}{24}$$

To **multiply** two or more fractions, multiply the numerators to obtain the numerator of the answer and multiply the denominators to obtain the denominator of the answer. Reduce the answer to lowest terms.

EXAMPLE 11.   Multiply: $5 \times \frac{4}{7}$

Solution: $5 \times \frac{4}{7} = \frac{5}{1} \times \frac{4}{7} = \frac{5 \times 4}{1 \times 7} = \frac{20}{7} = 2 \frac{6}{7}$

EXAMPLE 12.   Multiply: $\frac{4}{9} \times \frac{3}{8}$

Solution: $\frac{4}{9} \times \frac{3}{8} = \frac{4 \times 3}{9 \times 8} = \frac{12}{72} = \frac{1}{6}$

EXAMPLE 13. Multiply: $\dfrac{2}{3} \times \dfrac{1}{6} \times \dfrac{9}{2}$

Solution: $\dfrac{2}{3} \times \dfrac{1}{6} \times \dfrac{9}{2} = \dfrac{2 \times 1 \times 9}{3 \times 6 \times 2} = \dfrac{18}{36} = \dfrac{1}{2}$

In multiplication, **canceling** means dividing the numerator and the denominator of two different fractions, or the numerator and denominator of the same fraction, by the same nonzero number. The effect of canceling in a multiplication problem, is to reduce fractions to their lowest terms. Cancel whenever you have the opportunity, because it will make the numbers more manageable.

EXAMPLE 14. Multiply: $\dfrac{4}{5} \times \dfrac{7}{12}$

Solution: Dividing 4 in the numerator and 12 in the denominator by 4 gives:

$$\dfrac{4}{5} \times \dfrac{7}{12} = \dfrac{1}{5} \times \dfrac{7}{3} = \dfrac{1 \times 7}{3 \times 5} = \dfrac{7}{15}$$

The way that you "cancel" in a problem is not unique.

EXAMPLE 15. Multiply: $\dfrac{8}{12} \times \dfrac{2}{3} \times \dfrac{6}{9}$

Solution: $\dfrac{8}{12} \times \dfrac{2}{3} \times \dfrac{6}{9} = \dfrac{2}{1} \times \dfrac{2}{3} \times \dfrac{2}{9} = \dfrac{8}{27}$

When you are multiplying mixed numbers, such as 2 1/3 × 3 1/2, begin by converting them to improper fractions and proceed as before.

To **divide** fractions, multiply the first fraction by the reciprocal of the second fraction. Cancel when possible, and reduce the answer to lowest terms.

EXAMPLE 16. Divide: $\dfrac{5}{9} \div \dfrac{8}{3}$

Solution: $\dfrac{5}{9} \div \dfrac{8}{3} = \dfrac{5}{9} \times \dfrac{3}{8} = \dfrac{5}{3} \times \dfrac{1}{8} = \dfrac{5}{24}$

EXAMPLE 17. Divide: $\dfrac{7}{8} \div 2$

Solution: $\dfrac{7}{8} \div 2 = \dfrac{7}{8} \times \dfrac{1}{2} = \dfrac{7}{16}$

EXAMPLE 18. Divide: $1\dfrac{3}{5} \div 2\dfrac{2}{7}$

Solution: $1\dfrac{3}{5} \div 2\dfrac{2}{7}$ Begin by converting mixed numbers to improper fractions.

$\dfrac{8}{5} \div \dfrac{16}{7}$

$\dfrac{8}{5} \times \dfrac{7}{16} = \dfrac{1}{5} \times \dfrac{7}{2} = \dfrac{7}{10}$

2.2 DECIMALS

Numbers such as 7.6, $43.95, and 17.04 are called **decimal numbers**, or simply **decimals**. The decimal point "." in each number separates the whole number portion of the decimal from the fractional portion of the number.

$$7 \quad 8 \quad 2 \; . \; 8 \quad 0 \quad 5$$

     whole   fractional
     number   part

Decimals of the form 0.3, 0.75, and 0.625 are called **terminating decimals**. In a terminating decimal it is understood that the digit zero repeats indefinitely after the last nonzero digit. For example, 0.3 = 0.3000. . . , 0.75 = 0.7500. . . , and 0.625 = 0.625000. . . the three dots ". . ."means that the pattern of digits goes on forever in the indicated manner. A decimal that is not terminating is called a **nonterminating decimal**. Some nonterminating decimals are 0.777. . . , 5.27444. . . , 8.15115111511115. . . , and 0.393939. .

A decimal with a repeating digit or digits may be abbreviated by placing a bar " – " over the repeating digit or digits. Thus, 0.777. . . becomes 0.$\overline{7}$, 5.27444. . . becomes 5.27$\overline{4}$, and 0.393939. . . becomes 0.$\overline{39}$.

Two decimal numbers are equal if and only if they have the same digits relative to the decimal point. When two decimals are not equal, the larger may be determined by inspecting the place-value of the digits that differ in the two numbers.

EXAMPLE 19. If $63.47 = $6?.47, then the ? must equal?

Solution: It must be 3.

EXAMPLE 20. Is 7.402 < 7.432?

Solution: Moving left to right make a place by place inspection of the two decimals. Since "0" is less than "3" (and the other digits agree) 7.402 < 7.432.

To add decimals, line up the numbers (the **addends**) with all of the decimal points in a column, and begin to add in the column farthest to the right to obtain the **total**.

EXAMPLE 21. Add 39.66 feet + 7.02 feet + 197.45 feet.

Solution:
```
    39.66
     7.02
+ 197.45
  244.13 feet
```

To subtract decimals, line up the numbers with the decimal points in a column, and begin to subtract in the column farthest to the right.

EXAMPLE 22. How much is 92.60 minus 37.14?

Solution:
```
 92.60  ← This is called the minuend.
- 37.14 ← This is called the subtrahend.
 55.46  ← This is called the difference.
```

Remember, when the digit in the **subtrahend** is greater than the digit in the **minuend** you must "borrow" from the place to the left in your minuend. In Example 22, "1" had to be "borrowed" from the 9. When this "1" package was converted to 10 ones, we obtained a total of 12 ones, from which we were able to subtract 7 ones.

To multiply two decimals, ignore the decimal point for a moment and multiply the numbers. Point off, right to left, the number of decimal places equal to the sum of the decimal places in the two numbers.

EXAMPLE 23. Multiply 6.9 by 7.2.

Solution:

```
      6.9    1 decimal place in this multiplicand
    × 7.2    1 decimal place in this multiplier
    1 3 8
  4 8 3
  4 9.6 8.   1 + 1 = 2 decimal places to be
             pointed off in the product.
```

EXAMPLE 24. Multiply 51.7 × 3.44.

Solution:

```
      5 1.7
    ×   3.4 4
      2 0 6 8
    2 0 6 8
  1 5 5 1
  1 7 7.8 4 8.   1 + 2 = 3 decimal places to be
                 pointed off in the product.
```

To divide a decimal by a decimal, convert the division into an equivalent division with a whole number divisor. This is accomplished by multiplying both the numerator (called the **dividend**), and the denominator (called the **divisor**), by 10, 100, 1000, and so on. Use the power of 10 that will move the decimal point in the divisor all the way to the right. When this is done divide. The decimal point in the answer is directly above the decimal point in the dividend.

EXAMPLE 25. Evaluate $\dfrac{7.35}{3.5}$.

Solution: $\dfrac{7.35}{3.5} = \dfrac{7.35 \times 10}{3.5 \ \times 10} = \dfrac{73.5}{35} = 2.1$

$$
\begin{array}{r}
2.1 \\
3.5\ )\overline{7.3\ 5} \\
\underline{7\ 0} \\
3\ 5 \\
\underline{3\ 5} \\
0
\end{array}
$$

or 3.5)7.3 5   2.1 ← This is called the **quotient**.

0 ← This is called the **remainder**.

EXAMPLE 26. Divide 11.73 by 2.51.

Solution:

$$
\begin{array}{r}
4.6\ 7 \\
2.5\ 1.)\overline{1\ 1.7\ 3.0\ 0} \\
\underline{1\ 0\ 0\ 4} \\
1\ 6\ 9\ 0 \\
\underline{1\ 5\ 0\ 6} \\
1\ 8\ 4\ 0 \\
\underline{1\ 7\ 5\ 7} \\
8\ 3
\end{array}
$$

4.6 7 ← quotient

8 3 ← remainder

2.3 RATIO AND PROPORTION

A **ratio** is formed by writing a division of two numbers. The ratio of 5 automobiles to 15 automobiles is 5/15 which is equivalent to 1/3. The ratio of 7 to 8 may be expressed in these three ways:

1) By using a division sign, 7 ÷ 8.

2) By using the ratio sign " : ", 7:8.

3) By using a fraction, $\dfrac{7}{8}$.

A ratio is usually a comparison of two like quantities.

EXAMPLE 27. What is the ratio of 11 yards to 18 feet?

Solution: To compare like measures we must change yards to feet or feet to yards. If we change 11 yards to feet, the ratio is 33/18 which is equivalent to 11/6. If we change feet to yards, 18 feet becomes 6 yards and we obtain the same answer 11/6.

However, there are applied situations where one wishes to compare "unlike quantities" such as a ratio of miles traveled per gallon of gasoline.

EXAMPLE 28. In a certain school the student-teacher ratio is 30 to 1. Express this ratio in three ways.

Solution: This is a comparison of unlikes:

$$30 \div 1, \quad 30:1, \quad \frac{30}{1}$$

A **proportion** is an equation in which two ratios are equal. Some examples of proportions are:

$$\frac{3}{5} = \frac{12}{20} , \quad \frac{25}{200} = \frac{1}{8} , \quad \text{and} \quad \frac{12}{21} = \frac{20}{35}$$

A proportion is an equation of the form a/b = c/d, where b ≠ 0 and d ≠ 0. Sometimes it is convenient to write this in the form a:b = c:d. The four numbers are referred to as the "terms" of the proportion. The first and last terms, a and d, are called the **extremes**; the second and third terms, b and c, are called the **means**.

EXAMPLE 29. Given the proportion 4/5 = 20/25 state the means and the extremes.

Solution: The means are 5 and 20. The extremes are 4 and 25.

Rule for Proportions: In every proportion ,the product of the means, equals the product of the extremes. For a/b = c/d, a·d = b·c.

Frequently, applied problems arise in which one knows three of the four terms of the proportion and is asked to determine the missing term. Being asked to find the missing term is equivalent to being asked to "solve the proportion".

EXAMPLE 30. If $\dfrac{x}{5} = \dfrac{18}{30}$, find the value of x.

Solution: Using the Rule for Proportions we have:

$$30 \cdot x = 5 \cdot 18$$
$$30x = 90$$
$$x = 3$$

EXAMPLE 31. If $24:42 = 4:x$, find the value of x.

Solution: Using the Rule for Proportions we have:

$$24 \cdot x = 42 \cdot 4$$
$$24x = 168$$
$$x = 7$$

EXAMPLE 32. A driver knows that 1 inch on his map is equivalent to 40 miles. How many inches are equivalent to 180 miles?

Solution: We are being asked to solve this proportion:

$$\frac{1 \text{ inch}}{x \text{ inches}} = \frac{40 \text{ miles}}{180 \text{ miles}}$$

$$40 \cdot x = 180$$

$$x = 4 \ 1/2 \text{ inches}$$

2.4 EXPONENTS

Whenever two or more numbers are multiplied, each of the numbers is called a **factor**. For the case that all of the factors are the same such as $5 \times 5 \times 5$, $2 \times 2 \times 2 \times 2$, and 7×7 there is this shorthand notation: $5 \times 5 \times 5 = 5^3$, $2 \times 2 \times 2 \times 2 = 2^4$, and $7 \times 7 = 7^2$.

For any nonzero number "a" and for any natural number "n", $a^n = \underbrace{a \times a \times a \times \ldots \times a}_{n \text{ factors of } a}$.

In an expression of the form a^n it is customary to refer to the "n" as an **exponent** and the "a" as the **base**. For 3^5, 5 is the exponent and 3 is the base. The exponent tells us how many times to write the base down with multiplication understood. Sometimes a^n is called a **power**, and read "a to the nth power". Hence, 3^8 is read "three to the eighth power" and 2^4 is read "two to the fourth power".

To multiply exponential numbers having the same base, add the exponents. That is, $a^m \cdot a^n = a^{m+n}$.

EXAMPLE 33. Simplify $5^3 \cdot 5^4$.

Solution: $5^3 \cdot 5^4 = 5^{3+4} = 5^7$.

EXAMPLE 34. Simplify $8^6 \cdot 8^9$.

Solution: $8^6 \cdot 8^9 = 8^{6+9} = 8^{15}$.

To divide exponential numbers with the same base, subtract the exponents. That is, $a^m/a^n = a^{m-n}$.

EXAMPLE 35. Simplify $\dfrac{3^9}{3^5}$.

Solution: $\dfrac{3^9}{3^5} = 3^{9-5} = 3^4$

EXAMPLE 36. Simplify $\dfrac{7^6}{7}$.

Solution: 7(with no exponent) is shorthand for 7^1 so

$$\frac{7^6}{7} = 7^{6-1} = 7^5.$$

Clearly, if m = n, $a^m/a^n = 1$. Using the rule for dividing exponential numbers we obtain $a^m/a^n = a^{m-n} = a^0$ so a^0 must equal one. Mathematicians express this by saying "By definition for any nonzero number a, $a^0 = 1$". Thus, $8^0 = 1$, and $(-2)^0 = 1$.

If a is any nonzero number and n is any natural number, mathematicians define $a^{-n} = 1/a^n$.

EXAMPLE 37. Simplify 4^{-2}.

Solution: $4^{-2} = \dfrac{1}{4^2} = \dfrac{1}{16}$

EXAMPLE 38. Simplify $\dfrac{3^4}{3^6}$

Solution: $\dfrac{3^4}{3^6} = 3^{4-6} = 3^{-2} = \dfrac{1}{3^2} = \dfrac{1}{9}$

EXAMPLE 39. Simplify $5^{-6} \cdot 5^3$.

Solution: $5^{-6} \cdot 5^3 = 5^{-6+3} = 5^{-3} = \dfrac{1}{5^3} = \dfrac{1}{125}$

EXAMPLE 40. Simplify $4^7 \cdot 4^{-7}$.

Solution: $4^7 \cdot 4^{-7} = 4^{7-7} = 4^0 = 1.$

To find the power of a number expressed in exponential form, multiply the exponents. That is, $(a^m)^n = a^{mn}$ where m and n are **integers** (. . . -3, -2, -1, 0, 1, 2, 3, . . .).

EXAMPLE 41. Simplify $(4^3)^2$.

Solution: $(4^3)^2 = 4^{(3)(2)} = 4^6$

EXAMPLE 42. Simplify $(5^4)^{-1}$.

Solution: $(5^4)^{-1} = 5^{(4)(-1)} = 5^{-4} = \dfrac{1}{5^4}$

EXAMPLE 43. Simplify $(7^6)^0$.

Solution: $(7^6)^0 = 7^{(6)(0)} = 7^0 = 1$

EXAMPLE 44. Simplify $(2^{-2})^{-2}$.

Solution: $(2^{-2})^{-2} = 2^{(-2)(-2)} = 2^4 = 16$

2.5 FACTORING

The simplest form of **factoring** occurs when one "pulls out" a common monomial factor from each term of a polynomial. For example, $10x - 20$ contains the factor 10. We say that "10 is a common monomial factor of $10x - 20$". Using the distributive property one can write $10x - 20 = 10(x - 2)$. All three terms of the trinomial $a^3 - 7a^2b + 5a$ contain the common monomial factor a so one can write $a^3 - 7a^2b + 5a = a(a^2 - 7ab + 5)$. Do you see that $4xy$ is a common monomial factor of the expression $8x^2y^2 + 16xy^2 - 12xy$? In general, when one wishes to factor an algebraic expression, which means to write it as a product of prime expressions, the first step is to factor any common monomial.

EXAMPLE 45. Factor the common monomial from each of the following:

　　　　　a. $30x + 45$　　　　b. $x^3 - x$

　　　　　c. $a^2 + ab + b^2$　　d. $6c^2d^2 - 12c^2d - 9cd^2$

Solution: a. $15(2x + 3)$　　　　b. $x(x^2 - 1)$

　　　　c. There is none.　　d. $3cd(2cd - 4c - 3d)$

51

A binomial, such as $x^2 - 36$, is a difference of two terms each of which is a perfect square. By multiplying binomials one can readily verify that this difference of two perfect squares can be factored. $x^2 - 36 = (x + 6)(x - 6)$. Similarly, $9a^2 - 100 = (3a)^2 - 10^2 = (3a + 10)(3a - 10)$, and $5x^2 - 5y^2 = 5(x^2 - y^2) = 5(x + y)(x - y)$.

To factor a binomial that is the difference between two perfect squares, memorize this pattern: $a^2 - b^2 = (a + b)(a - b)$.

Binomials of the form $a^2 + b^2$ cannot be factored; they are prime. Thus, any binomial that is the sum of two perfect squares such as $x^2 + 16$, $m^2 + n^2$, and $16a^2 + 100$, cannot be factored.

EXAMPLE 46.   Factor: a. $x^2 - 25$

b. $4x^2 - 9y^2$

c. $am^2 - an^2$

Solution: a.   In fitting this expression to the pattern $a^2 - b^2$, x is playing the role of a and 5 is playing the role of b.

$$x^2 - 25 = (x)^2 - (5)^2 = (x + 5)(x - 5)$$

b.   In fitting this expression to the pattern $a^2 - b^2$, 2x is playing the roll of a and 3y is playing the role of b.

$$4x^2 - 9y^2 = (2x)^2 - (3y)^2 = (2x + 3y)(2x - 3y)$$

c.   You always begin a factoring problem by looking for a common monomial.

$am^2 - an^2 = a(m^2 - n^2)$;   a is the common monomial

$$= a(m + n)(m - n)$$

Sometimes, a problem calls for repeating the factoring pattern more than once. For example, $x^4 - y^4 = (x^2 - y^2)(x^2 + y^2)$. Now, while this factoring is correct, the first factor is another binomial difference between two perfect squares and the second factor is prime. Hence $x^4 - y^4 = (x + y)(x - y)(x^2 + y^2)$. Being asked to factor always means that you are to "completely factor" an expression. A completely factored expression has only prime factors.

A binomial of the form $x^3 - 64$ is the difference between two perfect cubes, x^3 and 4^3. The expression $a^3 + 1,000$ is the sum of two perfect cubes. Interestingly, one can factor both the difference of two perfect cubes and the sum of two perfect cubes. The discoverer of the following factoring patterns is anyone's guess.

> To factor a binomial that is the difference between two perfect cubes, or one that is the sum of two perfect cubes memorize these patterns:
> $$a^3 - b^3 = (a - b)(a^2 + ab + b^2)$$
> $$a^3 + b^3 = (a + b)(a^2 - ab + b^2)$$

Observe that each factorization is a product of a binomial and a trinomial. It may help you in memorizing these patterns to note that each factorization has exactly one negative sign. For the difference between two perfect cubes, it appears in the binomial factor. For the sum of two perfect cubes, it appears in the trinomial factor.

EXAMPLE 47. Factor: a. $x^3 - 64$

  b. $4a^3 - 32$

Solution: a. In fitting this expression to the pattern $a^3 - b^3$, x is playing the role of a and 4 is playing the role of b.

$$x^3 - 64 = (x)^3 - (4)^4 = (x - 4)(x^2 + 4x + 16)$$
$$\uparrow \qquad \uparrow$$
$$a \qquad b$$

  b. $4a^3 - 32 = 4(a^3 - 8)$; 4 is the common monomial
$$= 4(a - 2)(a^2 + 2a + 4)$$

53

EXAMPLE 48.   Factor: a. $64x^6 + 1$

b. $-2xy^3 - 54x$

Solution: a.   In fitting this expression to the pattern $a^3 + b^3$, $4x^2$ is playing the role of a, and 1 is playing the role of b.

$$64x^6 + 1 = \underbrace{(4x^2)}_{a}{}^3 + \underbrace{(1)}_{b}{}^3 = (4x^2 + 1)(16x^4 - 4x^2 + 1)$$

b. $-2xy^3 - 54x = -2x(y^3 + 27)$: $-2x$ is the common monomial

$$= -2x(y + 3)(y^2 - 3y + 9)$$

The trinomial $x^2 + 7x + 12$ can be factored into the product $(x + 3)(x + 4)$, and the trinomial $x^2 - 3x - 10$ can be factored into the product $(x - 5)(x + 2)$. To factor any trinomial of the form $x^2 + bx + c$, that is not prime, two things are important:

1)   the ability to multiply two binomials (Some people may refer to this as the "foil method of multiplying binomials").

and

2)   the ability to recognize the correct number combinations when you see them.

At first, recognizing the correct number combinations may require some "trial and error", but this recognition will improve rapidly as you practice factoring.

One attacks the problem of factoring $x^2 - 4x - 21$ as follows: The answer must be of the form $(x + ?)(x - ?)$. Here the two missing numbers must have a product of -21 and a sum of -4. Because the middle term is negative the negative number that we are seeking must have the greater absolute value of the two factors of -21. Multiply to verify that $x^2 - 4x - 21 = (x + 3)(x - 7)$.

To factor $y^2 + 9y + 18$ start with $(y + ?)(y + ?)$. Because the last term is 18, the two numbers we are seeking must be of like sign. Because the two numbers total 9 they must be positive.

54

Factors of 18	Product of Factors	Sum of Factors
1,18	18	19
2,9	18	11
3,6	18	9

Recalling how to multiply binomials, and maybe a little trial and error, tells us that $y^2 + 9y + 18 = (y + 6)(y + 3)$.

The trinomial $x^2 - 2x - 6$ is prime. The reader is encouraged to pick up their pencil and verify this fact.

EXAMPLE 49. Factor: a. $x^2 - 11x + 30$

b. $5a^2 + 85a - 90$

Solution: a. $x^2 - 11x + 30 = (x - 5)(x - 6)$

b. $5a^2 + 85a - 90 = 5(a^2 + 17a - 18)$

$= 5(a - 1)(a + 18)$

To factor a trinomial of the form $ax^2 + bx + c$ where a, b, and c are integers and a ≠ 0:

1. Find all possible pairs of binomials in which the product of the first terms is ax^2 and the product of the second terms is equal to c.

   a. If c is positive, the factors of c will be alike in sign (both + or both -) and the sign will be the sign of b.

   b. If c is negative, the factors of c will be unlike; one positive and the other negative.

2. Of these pairs, find the pair of binomials whose product yields the desired middle term; bx. If you exhaust all possible pairs without success your trinomial is prime.

3. Practice factoring. Factoring skill is developed through the experience of working many exercises.

EXAMPLE 50. Factor $3x^2 + 11x + 6$.

Solution: $3x^2 + 11x + 6 = (x + 3)(3x + 2)$

EXAMPLE 51. Factor $2x^2 - 9x - 5$.

Solution: $2x^2 - 9x - 5 = (2x + 1)(x - 5)$

EXAMPLE 52. Factor $14m + 8m^2 - 15$.

Solution: Begin by writing the problem as $8m^2 + 14m - 15$.

$$8m^2 + 14m - 15 = (4m - 3)(2m + 5)$$

EXAMPLE 53. Factor $-5 + 12t^2 + 28t$.

Solution: Begin by writing the problem as $12t^2 + 28t - 5$.

$$12t^2 + 28t - 5 = (6t - 1)(2t + 5)$$

EXAMPLE 54. Factor $8ab^2 + 2ab - 15a$.

Solution: $8ab^2 + 2ab - 15a = a(8b^2 + 2b - 15)$

$$= a(2b + 3)(4b - 5)$$

EXAMPLE 55. Factor $21x^2 - 2xy - 8y^2$.

Solution: $21x^2 - 2xy - 8y^2 = (7x + 4y)(3x - 2y)$

EXAMPLE 56. Factor $45a^3 - 63a^2b - 54ab^2$.

Solution: $45a^3 - 63a^2b - 54ab^2 = 9a(5a^2 - 7ab - 6b^2)$

$$= 9a(5a + 3b)(a - 2b)$$

2.6 LINEAR EQUATIONS

An **equation** is a statement that two quantities are equal. Here we are interested in equations with exactly one variable such as 6x = 12, -5y + 1 = 11, 3x + 2 = 4x - 1, and 1/2t = 17. A **linear equation** is one in which the exponent of the variable is one. Observe that this is true for all of the equations in our illustration.

The **solution** or the **root** of a linear equation is the number that when used in place of the variable (x, y, t and so on) makes the equation true. This is often expressed by saying "the solution satisfies the equation".

To solve an equation means to determine what value of your variable will satisfy the equation. An equation in the variable "x" is said to be solved when you can express it in the form "x" (meaning one x) equals some real number. Two equations that have the same solution are called **equivalent equations**. As an illustration:

$$7x - 3 = 25 \text{ is equivalent to}$$

$$7x = 28 \quad \text{and the solution or root is}$$

$$1 \cdot x \quad \text{or } x = 4$$

RULES FOR SOLVING EQUATIONS

1. Adding the same number to each side of an equation produces an equivalent equation.
2. Subtracting the same number from each side of an equation produces an equivalent equation.
3. Multiplying each side of an equation by the same nonzero number produces an equivalent equation.
4. Dividing each side of an equation by the same nonzero number produces an equivalent equation.

EXAMPLE 57. Solve the equation x - 9 = 12.

Solution: It is understood that you are to solve for "x".

$$x - 9 = 12$$

$$x - 9 + 9 = 12 + 9; \quad \text{add 9 to each side}$$

$$x + 0 = 21$$

$$x = 21$$

EXAMPLE 58. Solve the equation $\frac{y}{3}$ = 5.

Solution: It is understood that you are to solve for "y".

$$\frac{y}{3} = 12$$

$$(\frac{y}{3})(3) = (12)(3); \quad \text{multiply each side by 3}$$

$$y = 36$$

EXAMPLE 59. Solve the equation 4n + 6 = 26.

Solution:
$$4n + 6 = 26$$

$$4n + 6 - 6 = 26 - 6; \quad \text{subtract 6 from each side}$$

$$4n + 0 = 20$$

$$\frac{4n}{4} = \frac{20}{4} \quad ; \quad \text{divide each side by 4}$$

$$n = 5$$

To **check** an equation such as the one appearing in example 59, replace the variable in the equation (everywhere it appears) with the solution. If the solution is correct, working out the arithmetic on each side independently (work the left side and then work the right side) will yield the same number. Here is the check for example 59.

58

Check: Given 4n + 6   = 26

$$(4)(5) + 6 \; ? \; 26$$

$$26 = 26$$

EXAMPLE 60.   Solve the equation 5x + 3 - 2x = 18.

Solution: 5x + 3 - 2x   = 18

$$3x + 3 = 18; \quad \text{combine like terms}$$

$$3x + 3 + (-3) = 18 + (-3); \quad \text{add -3 to each side}$$

$$3x + 0 = 15$$

$$x = 5; \quad \text{divide each side by 3}$$

EXAMPLE 61.   Solve the equation -3(-2y + 1) = 12.

Solution: -3(-2y + 1) = 12

$$6y - 3 = 12; \quad \text{multiply each number in the parentheses by -3}$$

$$6y = 15; \quad \text{add 3 to each side}$$

$$y = 2\frac{1}{2}; \quad \text{divide each side by 6}$$

EXAMPLE 62.   Solve the equation 0.7y - 0.3 + 0.2y = 2.4.

Solution: 0.7y - 0.3 + 0.2y = 2.4

$$0.9y - 0.3 = 2.4; \quad \text{combine like terms}$$

$$0.9y = 2.7; \quad \text{add 0.3 to each side}$$

$$y = 3; \quad \text{divide each side by 0.9}$$

Check: (0.7)(3) - 0.3 + (0.2)(3) ? 2.4

$$2.1 - 0.3 + 0.6 \; ? \; 2.4$$

$$2.4 = 2.4$$

59

EXAMPLE 63. Solve the equation $\frac{-2x}{3} + 1 = -5$.

Solution: $\frac{-2x}{3} + 1 - 1 = -5 - 1$; subtract 1 from each side

$$\frac{-2x}{3} = -6$$

$$(-\frac{3}{2})(\frac{-2x}{3}) = (-\frac{3}{2})(-6); \quad \text{multiply each side by } -\frac{3}{2}$$

$$x = 9$$

2.7 LINEAR INEQUALITY

Given any two real numbers (or points) on the number line the number farther to the right is the larger. On the number line 5 is to the right of 3 and this is denoted either $5 > 3$, read "five is greater than 3", or $3 < 5$ read "three is less than five".   2 is to the right of -1.50, so $2 > -1.50$ or $-1.50 < 2$. -3.75 is to the right of -10, so $-3.75 > -10$ or $-10 < -3.75$.   The symbols "<" read "is less than", and the ">" read "is greater than" are called **inequality symbols**. An easy way to remember the meaning of the symbol is to notice that the wide part of the symbol is next to the larger number.

If you wish to say that "three is less than or equal to x", you can write $3 \leq x$. The symbol $\leq$ is an abbreviation of the two symbols < and =. When you write $x \geq 5$, this means that x is any real number greater than or equal to 5. That is, x can be 5, 6.2, 7.03, 8.88, 100 and so on.

EXAMPLE 64.   a.   Discuss the statement $y \leq 10$.

b.   Express the fact that t (for time) is greater than or equal to 30 minutes.

Solution: a. $y \leq 10$ means that y is less than or equal to 10.   Hence y is found to the left of 10 on the number line or at 10 itself.   Some possible values of y are: 9, 8.5, 0, -1, or -55.

b. $t \geq 30$ minutes.

A **linear inequality** in one variable is solved in exactly the same way that a linear equation in one variable is solved, with one noteworthy exception. When each side of an inequality is multiplied (or divided) by a negative number the direction of the inequality must be reversed. $6 > 5$ and $(6)(-3) < (5)(-3)$ or $-18 < -15$. If $-x \geq 1$, then $(-x)(-1) \leq (1)(-1)$ or $x \leq -1$.

EXAMPLE 65. Solve $7x - 8 < 13$

Solution: $7x - 8 \quad\quad < 13$

$\qquad 7x - 8 + 8 < 13 + 8$; add 8 to each side.

$\qquad\qquad 7x < 21$

$\qquad\qquad x < \ 3$; divide each side by 3.

This tells us that "x" is any real number on the number line to the left of 3. The numbers 2, 1.3, 0, -1/2, -7, and so on satisfy this inequality.

EXAMPLE 66. Solve $\frac{3}{2} x \geq 9 + 2x$

Solution: $\frac{3}{2} x \quad\quad \geq 9 + 2x$

$\qquad \frac{3}{2} x - 2x \geq 9 + 2x - 2x$; add -2x to each side

$\qquad -\frac{1}{2} x \geq 9$; combine like terms

$\qquad\qquad x \leq -18$; multiply each side by -2

EXAMPLE 67. Solve $0.4(15m - 16) > 12$

Solution: $0.4(15m - 6) \quad > 12$

$\qquad\qquad 6m - 2.4 > 12$; distribute 0.4

$\qquad 6m - 2.4 + 2.4 > 12. + 2.4$; add 2.4 to each side

$\qquad\qquad 6m > 14.4$

$\qquad\qquad m > 2.4$; divide each side by 6

EXAMPLE 68. Solve $-\frac{1}{3}t + 7 \leq \frac{3}{4}$

Solution: $-\frac{1}{3}t + 7 \qquad \leq \frac{3}{4}$

$\qquad -4t + 84 \qquad \leq 9;$ clear the inequality of fractions
$\qquad\qquad\qquad\qquad\qquad$ multiply by 12

$\qquad -4t + 84 - 84 \leq 9 - 84;$ subtract 84 from each side

$\qquad\qquad -4t \leq -75$

$\qquad\qquad t \geq 18\frac{3}{4}$; divide each side by -4

2.8   EVALUATING EXPRESSIONS

To evaluate a given expression substitute the given values for the indicated unknowns.

EXAMPLE 69.   Given the equation $y = mx + b$, determine the value of y when m = 1/2, x = 6 and b = -1.

Solution: $y = mx + b$

$\qquad y = (\frac{1}{2})(6) + (-1)$

$\qquad y = 3 + (-1)$

$\qquad y = 2$

EXAMPLE 70.   Given the equation $y = mx + b$, determine the value of x when y = 23, m = 5 and b = 3.

Solution: $y = mx + b$

$\qquad 23 = 5x + 3$

$\qquad 20 = 5x$

$\qquad 4 = x$

EXAMPLE 71. Given the equation $a^2 + b^2 = c^2$, determine the value of c when a = 5 and b = 12.

Solution: $c^2 = a^2 + b^2$

$c^2 = 5^2 + 12^2$

$c^2 = 25 + 144$

$c = \sqrt{25 + 144}$

$c = \sqrt{169} = 13$

EXAMPLE 72. The volume of a sphere is $V = 4/3 \cdot \pi \cdot R^3$ where π is approximately 3.14. For the case that R = 2 inches express the volume. If you have a calculator express the answer correct to the nearest hundredth.

Solution: $V = \left(\dfrac{4}{3} \right)(3.14)(2^3) = 33.49$ cubic inches.

EXAMPLE 73. The formula C = 5(F − 32)/9 is used to convert temperatures from the Fahrenheit scale to the Celsius scale. When F = 212°, what is the value of C?

Solution: $C = \dfrac{5(212 - 32)}{9} = \dfrac{5(180)}{9} = 100°$

EXAMPLE 74. If g = 0.2, h = 2.5, and k = 4.2, what is the value of the expression (3k − 2h)/g?

Solution: $\dfrac{3k - 2h}{g} = \dfrac{(3)(4.2) - (2)(2.5)}{0.2}$

$= \dfrac{12.6 - 5}{0.2}$

$= \dfrac{7.6}{0.2}$

$= 38$

2.9 FRACTIONAL EQUATIONS

An equation in which the variable appears in the denominator of at least one term is called a **fractional equation**. When both sides of a fractional equation are multipled by the least common denominator (LCD) of all terms, a new equation called the **derived equation** is obtained. Some examples of fractional equations are:

$$\frac{1}{3} - \frac{2}{x} = 5; \quad \frac{1}{2y} + \frac{2}{y} = \frac{1}{3} ; \text{ and } \quad \frac{5n}{n-1} + 4 = \frac{6}{n-2}$$

To Solve Fractional Equations

1. Find the LCD of all fractions in the equation.
2. Multiply each side of the fractional equation by the LCD. Be sure that you have multiplied all of the terms on both sides by the LCD.
3. Solve the derived equation.
4. Substitute the solution for the variable in the given equation. If the solution does not satisfy the given equation it is an "extraneous solution".

EXAMPLE 75. Solve $\frac{2}{3x} = \frac{1}{9}$

Solution: Step 1: The LCD of 3x and 9 is 9x.

Step 2: $9x\left(\frac{2}{3x} \right) = 9x\left(\frac{1}{9} \right)$

Step 3: $3(2) = x(1)$

$6 = x$

Step 4: $\frac{2}{(3)(6)} \; ? \; \frac{1}{9}$

$\frac{1}{9} = \frac{1}{9}$

EXAMPLE 76. Solve $\dfrac{5}{n} - \dfrac{1}{2} = 2$

Solution: Step 1:   The LCD of n and 2 is 2n.

Step 2: $2n\left(\dfrac{5}{n}\right) - 2n\left(\dfrac{1}{2}\right) = 2n(2)$

Step 3: $2(5) - n(1) = 4n$

$10 - n = 4n$

$10 = 5n$

$2 = n$

Check: Step 4: $\dfrac{5}{2} - \dfrac{1}{2} \; ? \; 2$

$\dfrac{4}{2} \; ? \; 2$

$2 = 2$

EXAMPLE 77. Solve $\dfrac{n - 2}{n + 3} = \dfrac{3}{8}$

Solution: $8(n + 3)\left(\dfrac{n - 2}{n + 3}\right) = 8(n + 3)\left(\dfrac{3}{8}\right)$

$8(n - 2) = (n + 3)(3)$

$8n - 16 = 3n + 9$

$5n = 25$

$n = 5$

EXAMPLE 78. Solve $\dfrac{x}{x - 1} = \dfrac{8}{x + 2}$

Solution: $(x - 1)(x + 2)\left(\dfrac{x}{x - 1} \right) = (x - 1)(x + 2)\left(\dfrac{8}{x + 2} \right)$

$$(x + 2)(x) = (x - 1)(8)$$

$$x^2 + 2x = 8x - 8$$

$$x^2 - 6x + 8 = 0$$

$$(x - 4)(x - 2) = 0$$

Answer: $x = 4$ and $x = 2$

2.10 STRAIGHT LINES

The "steepness" of a straight line is measured by comparing the vertical change to the horizontal change as a point moves along the line. The ratio of vertical change to horizontal change is called the **slope** of the line denoted by the letter "m".

More formally if one starts at point $P_1(x_1,y_1)$, read "the point P sub-one whose coordinates are x-one and y-one", and moves to the point $P_2(x_2,y_2)$, read "the point P sub-two whose coordinates are x-two and y-two", the slope may be expressed:

$$m = \text{the slope} = \frac{y_2 - y_1}{x_2 - x_1} ; \text{ where } x_2 \neq x_1.$$

Some people prefer the notation Δx read "delta x" to denote $x_2 - x_1$, and Δy read "delta y" to denote $y_2 - y_1$.

EXAMPLE 79. If $P_1(-1,2)$ and $P_2(3,4)$, determine Δy and Δx.

Solution: $\Delta y = y_2 - y_1 = 4 - 2 = 2$

$\Delta x = x_2 - x_1 = 3 - (-1) = 4$

EXAMPLE 80. Find the slope of the line passing through the points $P_1(3,1)$ and $P_2(5,2)$.

Solution: $m = \dfrac{y_2 - y_1}{x_2 - x_1} = \dfrac{2 - 1}{5 - 3} = \dfrac{1}{2}$

EXAMPLE 81. Find the slope of the line passing through the points $P_1(-1,4)$ and $P_2(2,-2)$.

Solution: $m = \dfrac{y_2 - y_1}{x_2 - x_1} = \dfrac{-2 - 4}{2 - (-1)} = \dfrac{-6}{3} = -2$

EXAMPLE 82. Find the slope of the line passing through $P_1(-2,5)$ and $P_2(3,5)$.

Solution: $m = \dfrac{5 - 5}{3 - (-2)} = \dfrac{0}{5} = 0$

    As the figure below illustrates some lines have positive slopes, and some lines have negative slopes. For lines parallel to the X-axis the slope is zero, and for lines parallel to the Y-axis the slope is undefined. The slope is undefined or meaningless when the denominator of m is zero. In mathematics division by zero is meaningless.

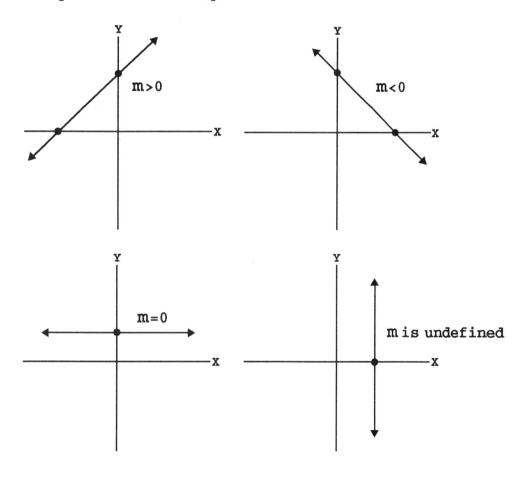

A straight line is represented by a linear equation in two variables. $x + y = 5$, $-2x + 4y = 1$, $y = -x + 5$, and $y = 2/3x + 1/2$ are equations of straight lines. Sometimes the second variable in the equation is not apparent. For example, $5x = 7$ is an abbreviation of $5x + 0y = 7$, and $-3y = 2$ is an abbreviation of $0x - 3y = 2$.

A linear equation solved for y, expressed in the form $y = mx + b$, is said to be expressed in **slope-intercept form**. When the equation of a straight line is solved for y the coefficient of x "m" is the slope of the line, and the second term denoted "b" is the y portion of the y-intercept. Solving a straight line equation for y allows one to immediately read off valuable information about the line. Given $y = -2x + 6$, the slope is -2 and the y-intercept is (0,6).

EXAMPLE 83. Given the equation $2x - y = 8$ determine the slope and the y-intercept of this line.

Solution: When this equation is expressed in slope-intercept form, $y = my + b$, the desired information is apparent. Solve this equation for y.

$$2x - y = 8$$

$$- y = -2x + 8$$

$$y = 2x - 8$$

Answer: The slope is $m = 2$ and the y-intercept is $(0,-8)$.

EXAMPLE 84. Find the equation of the line whose slope is 4 and whose y-intercept is (0,5).

Solution: $m = 4$ and $b = 5$. Substituting these values in the slope-intercept form $y = mx + b$ gives $y = 4x + 5$.

EXAMPLE 85. Find the equation of the line whose slope is $-1/3$ and whose y-intercept is (0,4).

Solution: $m = -\frac{1}{3}$, $b = 4$ so $y = -\frac{1}{3}x + 4$

EXAMPLE 86. Determine the equation of the line passing through the point (3,5) with a slope of -1.

Solution: We are given that m = -1. Placing this value in the slope-intercept form we obtain: y = -x + b. Next, the given point (3,5) must satisfy this equation. Hence:

$$y = -x + b$$

$$5 = -3 + b$$

$$b = 8$$

Answer: y = -x + 8

EXAMPLE 87. Determine the equation of the line passing through the point (8,6) with a slope of 2/3.

Solution: m = 2/3 so we have y = (2/3)x + b. Replacing y by 8 and x by 6 we have:

$$8 = \left(\frac{2}{3}\right)(6) + b$$

$$b = 4$$

Answer: $y = \frac{2}{3}x + 4$

EXAMPLE 88. Determine the equation of the line passing through the points (-2,5) and (4,6).

Solution: $m = \frac{6 - 5}{4 - (-2)} = \frac{1}{6}$ so $y = \frac{1}{6}x + b$.

Replacing y by 6 and x by 4 we have:

$$6 = \left(\frac{1}{6}\right)(4) + b$$

$$6 = \frac{2}{3} + b$$

$$b = 5\frac{1}{3}$$

Answer: $y = \frac{1}{6}x + 5\frac{1}{3}$

EXAMPLE 89. Determine the equation of the line passing through (6,5) and parallel to the X-axis.

Solution: All lines parallel to the X-axis have a slope of zero, so $y = 0x + b$ or $y = b$. Substituting $x = 6$ and $y = 5$ in this equation gives us $y = 5$.

Answer: $y = 0x + 5$ or $y = 5$

Two points uniquely determine a straight line. Hence, one way to graph a straight line (but not the only way), such as $2x + 6y = 12$, is to determine two points. If we substitute $x = 1$, in this equation, and solve for y, we obtain $y = 5/3$. This gives us the point $(1,5/3)$. When we substitute $x = 2$, we obtain the point $(2,4/3)$, and when we substitute $x = 3$, we obtain the point $(3,1)$. One can substitute a negative number for x, and one can substitute a fraction for x. However, for lines that are neither horizontal nor vertical nor pass through the origin, two points are readily determined. These are the points where the line crosses the X-axis and the Y-axis; they are called the **x-intercept** and the **y-intercept**, respectively.

Given the equation $ax + by = c$ where a and b are not zero:
1. To determine the x-intercept, let $y = 0$ and and solve the equation for x. The x-intercept is always of the form (a real no., 0).
2. To determine the y-intercept, let $x = 0$ and and solve the equation for y. The y-intercept is always of the form (0, a real no.).

EXAMPLE 90. Graph the equation $2x + 6y = 12$.

Solution: Let $x = 0$, then $2(0) + 6y = 12$

$$6y = 12$$

$$y = 2$$

The y-intercept is $(0,2)$.

Let y = 0, then 2x + 6(0) = 12

$$2x = 12$$

$$x = 6$$

The x-intercept is (6,0).

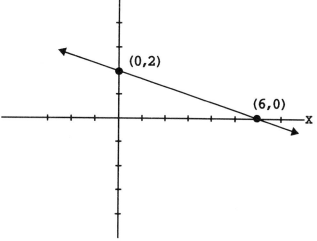

EXAMPLE 91.   Graph the equation y = 3x.

Solution: If x = 0, then y = 0.   In this case the two intercepts
are the same point, namely the origin.   Here, one must
find another point.   If x = 1, then y = 3, giving us
the point (1,3).

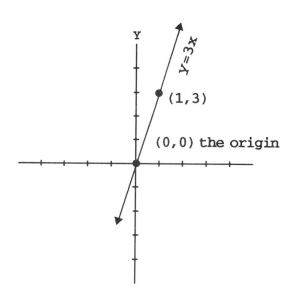

EXAMPLE 92.   Graph the equation x = 2.

Solution: x = 2 is an abbreviation of x + 0y = 2.   For all values
         of y, x = 2.   This straight line passes through (2,0)
         and is parallel to the Y-axis.

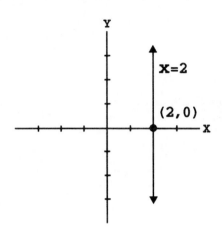

2.11 **RADICALS**

     The **square root** of a nonnegative number "b" is one of its
two equal factors.   Hence, the square root of b is a number a
such that a^2 = b.   A positive number such as 25 has two square
roots +5 and −5.   The numbers +5 and −5 are called the **positive
square root** and the **negative square root**, respectively.   The
square roots of 64 are +8 and −8.   Notice 5^2 = 25, $(-5)^2$ = 25,
8^2 = 64, and $(-8)^2$ =64.   Zero, 0, has only one square root,
itself.   A negative number, such as −9, has no square root
because there is no number "a" such that a^2 = −9.

     We refer to the positive square of a number as its **principal
square root**.   To denote the principal square root of a number
place a **radical sign** ($\sqrt{\ }$) over the number.   Some principal
square roots are: $\sqrt{100}$ = 10, $\sqrt{49}$ = 7, and $\sqrt{1/4}$ = 1/2.   The
number inside the radical sign is called the **radicand**.   In our
illustration the radicands are: 100, 49 and 1/4.

EXAMPLE 93.   a.   Find the square roots of 144.

              b.   Find $\sqrt{144}$.

Solution: a. Here the request is to find both square roots of
             144.   The answer is +12 and −12.

           b. Here the request is to find the principal square
              root.   The answer is 12.

72

EXAMPLE 94.　a.　Find the square roots of 4.

　　　　　　　　b.　Find $\sqrt{4}$.

　　　　　　　　c.　Find $-\sqrt{4}$.

Solution:　a. +2 and -2

　　　　　b. 2

　　　　　c. -2

EXAMPLE 95.　Solve the equation $x^2 = 36$.

Solution: Here we are being asked to find all of the values of x that will satisfy this equation.　The answer is both square roots of 36.　That is, $x = \pm \sqrt{36} = \pm 6$.

The opposite of cubing a number is taking its cube root. The **cube root** of a number "b" is denoted $\sqrt[3]{b}$ and $\sqrt[3]{b} = a$ if and only if $a^3 = b$.　Every number has exactly one cube root.　As an illustration $\sqrt[3]{27} = 3$, $\sqrt[3]{-8} = -2$, $\sqrt[3]{0} = 0$, and $\sqrt[3]{0.027} = 0.3$.

EXAMPLE 96.　Solve the equation $t^3 - 216 = 0$.

Solution: $t^3 - 216 = 0$

　　　　　　　　$t^3 = 216$;　add 216 to each side

　　　　　　　　$t = \sqrt[3]{216}$;　take the cube root of each side

　　　　　　　　$t = 6$

Using the **Product Rule for Radicals**, $\sqrt[n]{x}\ \sqrt[n]{y} = \sqrt[n]{xy}$ where $x \geq 0$ and $y \geq 0$, one can multiply radicals and simplify radicals. When n is not written it is understood to be 2.　As an illustration: $\sqrt{7}\ \sqrt{3} = \sqrt{21}$, $\sqrt{8}\ \sqrt{10} = \sqrt{80}$, and $\sqrt[3]{2}\ \sqrt[3]{6} = \sqrt[3]{12}$. Further $\sqrt{20} = \sqrt{4}\ \sqrt{5} = 2\ \sqrt{5}$, $\sqrt{72} = \sqrt{36}\ \sqrt{2} = 6\ \sqrt{2}$, and $\sqrt[3]{128} = \sqrt[3]{64}\ \sqrt[3]{2} = 4\ \sqrt[3]{2}$.

EXAMPLE 97.   a.   Multiply $\sqrt{5}\ \sqrt{6}$.

              b.   Simplify $\sqrt{300}$.

Solution: a. $\sqrt{5}\ \sqrt{6} = \sqrt{30}$

             b. $\sqrt{300} = \sqrt{100}\ \sqrt{3} = 10\ \sqrt{3}$

EXAMPLE 98.   Solve the equation $x^4 = 48$.

Solution: Here we are being asked to find all the values of x that satisfy this equation. The answer is:

$$x = \sqrt[4]{48} = \sqrt[4]{16}\ \sqrt[4]{3} = 2\ \sqrt[4]{3} \text{ and}$$

$$x = -\sqrt[4]{48} = -\sqrt[4]{16}\ \sqrt[4]{3} = -2\ \sqrt[4]{3}.$$

To Solve Equations with a Single Radical

1. Isolate the radical term on one side of the equation.
2. Square each side of the equation.
3. Solve the resulting equation.
4. Check each solution by substituting it in the original radical equation, being careful to take the principal square root of the number in the radicand.

EXAMPLE 99.   Solve $\sqrt{3x - 2} = 5$.

Solution: Step 1. $\sqrt{3x - 2} = 5$

         Step 2. $(\sqrt{3x - 2})^2 = 25$

                      $3x - 2 = 25$

         Step. 3. $3x = 27$

                          $x = 9$

Check: Step 4. $\sqrt{3(9) - 2}$? 5

                      $\sqrt{25}$? 5

                         $5 = 5$

When one solves a radical equation the algebra sometimes yields a solution that does not satisfy the original radical equation. A solution such as this, called an **extraneous solution**, is not part of the answer. This is illustrated in the next example.

EXAMPLE 100. Solve the equation $3 = 2\sqrt{x} + x$.

Solution: Step 1.
$$3 = 2\sqrt{x} + x$$
$$3 - x = 2\sqrt{x}$$

Step 2.
$$(3 - x)^2 = (2\sqrt{x})^2$$
$$9 - 6x + x^2 = 4x$$

Step 3.
$$x^2 - 10x + 9 = 0$$
$$(x - 1)(x - 9) = 0$$
$$\text{Either } x - 1 = 0 \quad \text{or} \quad x - 9 = 0$$
$$x = 1 \quad \text{or} \quad x = 9$$

We have two "potential solutions"
$$x = 1 \quad \text{and} \quad x = 9$$

Check: Step 4.

Can $x = 1$?	Can $x = 9$?
$3 \; ? \; 2\sqrt{1} + 1$	$3 \; ? \; 2\sqrt{9} + 9$
$3 \; ? \; 2 + 1$	$3 \; ? \; 2(3) + 9$
$3 = 3$ Yes.	$3 \neq 15$ No. This is an extraneous solution.

2.12 QUADRATIC EQUATIONS

Equations in one variable, call it x, in which the x is squared are called **quadratic equations** in one variable. We are interested in equations such as: $x^2 = 121$, $-3x^2 + 1 = 0$, $x^2 - x - 6 = 0$, and $(x + 5)^2 = 64$.

If $ab = 0$, then at least one of the factors a or b must be zero. This is the **Principle of a Zero Product**. The first way that one attempts to solve a quadratic equation is to set it equal to zero and then to factor the nonzero side. When this can be done we use the Principle of a Zero Product to obtain the two "potential" solutions of the quadratic equation.

EXAMPLE 101. Solve the equation $7x^2 - 21x = 0$.

Solution: We begin by noticing that the equation is already set equal to zero. This is a necessary first step.

$$7x^2 - 21x = 0$$

$$7x(x - 3) = 0; \quad \text{factor the nonzero side}$$

Using the Principle of a Zero Product

Either $7x = 0$ or $x - 3 = 0$

$$x = 0 \qquad\qquad x = 3$$

EXAMPLE 102. Solve the equation $n^2 + 2n = 35$.

Solution: $n^2 + 2n \qquad = 35$

$n^2 + 2n - 35 \quad = \quad 0; \text{ subtract 35 from each side}$

$(n + 7)(n - 5) = 0; \text{ factor the nonzero side}$

Either $\quad n + 7 = 0$ or $\quad n - 5 = 0$

$$n = -7 \qquad\qquad n = 5$$

EXAMPLE 103. Solve the equation $3y^2 = 5y + 12$.

Solution: $3y^2 \qquad\qquad = 5y + 12$

$3y^2 - 5y - 12 \quad = 0; \quad \text{add } -5y \text{ and } -12 \text{ to each side}$

$(3y + 4)(y - 3) = 0; \quad \text{factor the nonzero side}$

Either $3y + 4 \quad = 0$ or $\quad y - 3 = 0$

$$3y \quad = -4 \qquad\qquad y = 3$$

$$y \quad = -\frac{4}{3}$$

What does one do with a quadratic equation, such as $x^2 + x + 1 = 0$, which cannot be factored? An equation of the form $ax^2 + bx + c = 0$, where $a \neq 0$, is expressed in **standard form**.

$$4x^2 - 7x + 8 = 0 \qquad \text{or} \qquad 4x^2 + (-7)x + 8 = 0$$

$$\underset{a \qquad b \qquad c}{\uparrow \qquad \uparrow \qquad \uparrow} \qquad\qquad \underset{a \qquad b \qquad c}{\uparrow \qquad \uparrow \qquad \uparrow}$$

To express a quadratic equation in standard form, begin by setting the terms of the equation equal to 0. It is convenient to make the coefficient of x^2 a positive number. We will follow this suggestion.

EXAMPLE 104. Express the equation $3t^2 = 6t - 14$ in standard form and determine the values of a, b, and c.

Solution: $3t^2 - 6t + 14 = 0$

$$a = 3, \quad b = -6 \quad \text{and} \quad c = 14$$

For the case in which the standard form of a quadratic equation cannot be factored, for example $3x^2 + 8x - 8 = 0$, mathematicians have derived the **quadratic formula**.

THE QUADRATIC FORMULA

The solutions of the equation $ax^2 + bx + c = 0$ are:

$$x = \frac{-b \pm \sqrt{b^2 - 4ac}}{2a} \; ; \; a \neq 0$$

EXAMPLE 105. Solve the equation $y^2 + 2y = 8$ using the quadratic formula.

Solution: $a = 1$, $b = 2$, and $c = -8$. Substitute these values into the quadratic formula to obtain:

$$y = \frac{-2 \pm \sqrt{2^2 - 4(1)(-8)}}{2(1)}$$

$$y = \frac{-2 \pm \sqrt{4 + 32}}{2} = \frac{-2 \pm \sqrt{36}}{2}$$

$$y = \frac{-2 \pm 6}{2} \begin{cases} \dfrac{-2 + 6}{2} = 2 \\[2mm] \dfrac{-2 - 6}{2} = -4 \end{cases}$$

EXAMPLE 106. Solve the equation $x^2 - 10x + 25 = 0$.

Solution: $a = 1$, $b = -10$, and $c = 25$

$$x = \frac{-(-10) \pm \sqrt{(-10)^2 - 4(1)(25)}}{2}$$

$$x = \frac{10 \pm \sqrt{100 - 100}}{2} = \frac{10 \pm \sqrt{0}}{2} = 5$$

There are two solutions. They are both $x = 5$. Sometimes this is called a "double solution" of 5.

EXAMPLE 107. Solve the equation $2x^2 = -x + 1$.

Solution: $a = 2$, $b = 1$, and $c = -1$

$$x = \frac{-1 \pm \sqrt{1^2 - (4)(2)(-1)}}{2(2)}$$

$$x = \frac{-1 \pm \sqrt{1 + 8}}{4} = \frac{-1 \pm \sqrt{9}}{4}$$

Answers:

$$x = \begin{cases} \dfrac{-1 + 3}{4} = \dfrac{1}{2} \\[2mm] \dfrac{-1 - 3}{4} = -1 \end{cases}$$

EXAMPLE 108. Solve the equation $3x^2 + 8x - 8 = 0$

Solution: $a = 3$, $b = 8$, and $c = -8$

$$x = \frac{-8 \pm \sqrt{8^2 - (4)(3)(-8)}}{2(3)} = \frac{-8 \pm \sqrt{160}}{6}$$

$$x = \frac{-8 \pm 4\sqrt{10}}{6} = \frac{2(-4 \pm 2\sqrt{10})}{6}$$

Answer: $x = \dfrac{-4 + 2\sqrt{10}}{3}$ and $x = \dfrac{-4 - 2\sqrt{10}}{3}$

The second most common equation appearing in textbooks (after the equation of the straight line) is the equation of a parabola. A **parabola**, in "x", is an equation of the form $y = ax^2 + bx + c$ where $a \neq 0$. When $a < 0$ the graph of the equation looks like a "smooth symmetric mountain" opening forever downward and when $a > 0$ the graph of the equation looks like a "smooth symmetric valley" between two mountains opening forever upward. When the graph opens upward we say it is **concave up**. When the graph opens downward we say it is **concave down**. The parabola $y = x^2 + 1$ is graphed below.

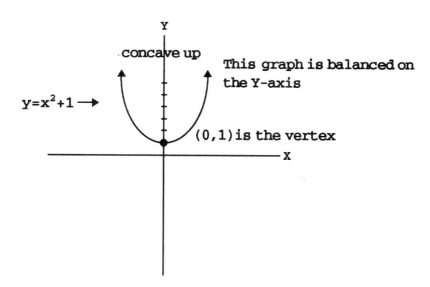

Every parabola in "x" is balanced upon the Y-axis or a line parallel to the Y-axis. This line is called the **line of symmetry**. The topmost point or the bottommost point of the parabola is called the **vertex**.

Parabolas of the form y = ax² + bx + c where a ≠ 0.

1. If a < 0 the parabola is concave down. If a > 0 the parabola is concave up.
2. The line of symmetry is parallel to the Y-axis. Its equation is x = -b/2a.
3. The parabola has exactly one vertex and its x coordinate is x = -b/2a. Substituting this value of x in the equation of the parabola gives the y coordinate of the vertex.

EXAMPLE 109. Graph the parabola y = x² - 3.

Solution: Recall this is an abbreviation of y = x² + 0x - 3.
a = 1, b = 0, and c = -3.

1. Because a = 1 > 0 the graph is concave up.

2. The line of symmetry is

$$x = -\frac{b}{2a} = -\frac{0}{2(1)} = 0$$

3. The x coordinate of the vertex is 0. The y coordinate is y = 0² - 3 = -3, so V(0,-3).

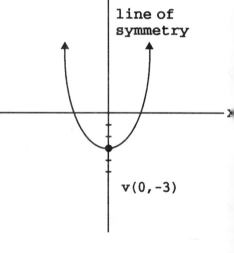

EXAMPLE 110. Discuss the graph of the parabola y = -x² + 4x - 2.

Solution: 1. The curve is concave down.

2. The equation of the line of symmetry is

$$x = -\frac{b}{2a} = -\frac{4}{2(-1)} = 2$$

3. The x coordinate of the vertex is also x = 2. Therefore, y = -2² + 4(2) - 2 = 2. The vertex is (2,2).

80

2.13 EXPONENTIAL EQUATIONS

Consider the following table and graph of $y = 2^x$ for some value of x.

x	-3	-2	-1	0	1/2	1	3/2	2	5/2
$y = 2^x$	1/8	1/4	1/2	1	1.4	2	2.8	4	5.7

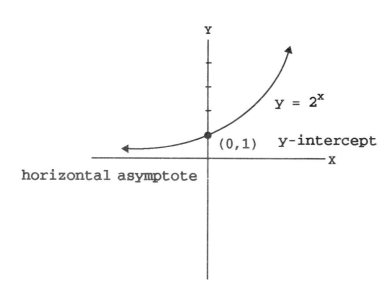

All equations of the form $y = a^x$ where $a > 0$ and $a \neq 1$, pass through the point (0,1). That is, (0,1) is the y-intercept for each equation. The graph of each equation lies above the X-axis (in quadrant I and in quadrant II). The X-axis is a horizontal asymptote. Further:

Moving left to right along the X-axis	Illustration
when $a > 1$, $y = a^x$ is increasing	$y = 3^x$
$\quad 0 < a < 1$, $y = a^{-x}$ is increasing	$y = (1/2)^{-x}$
when $a > 1$, $y = a^{-x}$ is decreasing	$y = 3^{-x}$
$\quad 0 < a < 1$, $y = a^x$ is decreasing	$y = (1/3)^x$

If your instructor refers to the "exponential function", he or she is referring to the equation $y = e^x$ (e is approximately 2.72).

EXAMPLE 111. Graph the equation $y = 5^{-x}$.

Solution: The y-intercept is $(0,1)$. The x-axis is a horizontal asymptote and the graph is steadily decreasing.

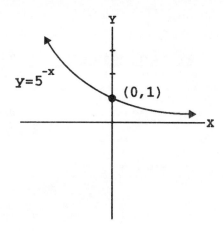

EXAMPLE 112. Graph the equation $y = e^x$.

Solution: The y-intercept is $(0,1)$. The x-axis is a horizontal asymptote and the graph is steadily increasing.

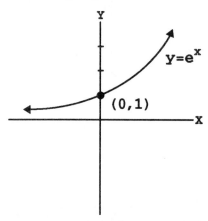

Because the exponential expression a^x yields a unique value for each value of x, $a^m = a^n$ if and only if $m = n$.

When the variable in the equation appears as part of the exponent it is called an **exponential equation**. For example, $7^{3x + 1} = 7^2$ is an exponential equation. Because the exponential expressions in the equation are to the same base, this equation is solved by equating the exponents. Thus $3x + 1 = 2$ which implies that $x = 1/3$.

EXAMPLE 113.   Solve the equation $6^{x+5} = 6^4$.

Solution: The bases are the same, so equate the exponents.

$$x + 5 = 4$$
$$x = -1$$

For the case that the exponential expressions are to different bases, convert all of the expression to the same base and then equate the exponents.

EXAMPLE 114.   Solve the equation $4^{5x+2} = 16^3$.

Solution: To obtain the same base, recall that $16 = 4^2$.

$$4^{5x+2} = 16^3 = (4^2)^3 = 4^6 \text{ so}$$

$$4^{5x+2} = 4^6$$

$$5x + 2 = 6; \quad \text{equate the exponents}$$

$$5x = 4$$

$$x = \frac{4}{5}$$

EXAMPLE 115.   Solve the equation $a^{2x+1} = a^{x-7}$.

Solution: $2x + 1 = x - 7;$ equate the exponents

$$x = -8$$

EXAMPLE 116.   Solve the equation $2^{x^2-1} = 64$

Solution: $2^{x^2-1} = 2^6; \quad 64 = 2^6$

$$x^2 - 1 = 6; \quad \text{equate the exponents}$$

$$x^2 = 7$$

$$x = \pm\sqrt{7}$$

2.14 LOGARITHMS

The **logarithm**, abbreviated "log", of a positive number is an exponent to a specified base. The logarithm of 64 to the base 4 is expressed $\log_4 64$. To determine $\log_4 64 = y$ is equivalent to solving the exponential equation $64 = 4^y$. Because $64 = 4^3$, the value of y is 3. Thus $\log_4 64 = 3$.

> The **logarithm of x > 0** to the base a is defined
> as follows: $\log_a x = y$ if and only if $x = a^y$
> where $a > 0$ and $a \neq 1$.

The preceding definition tells us that "log" and exponential statements occur in pairs. To each log statement there is a corresponding exponential statement and to each exponential statement there is a corresponding log statement. Use the following pair of equations as a memory device.

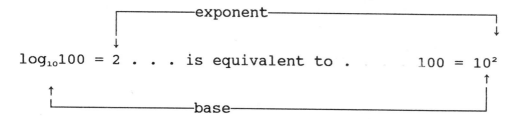

$$\log_{10} 100 = 2 \quad \ldots \text{ is equivalent to .} \quad 100 = 10^2$$

EXAMPLE 117. Find $\log_5 25$.

Solution: $\log_5 25 = y$ if and only if $25 = 5^y$ but $25 = 5^2$ so $y = 2$ and $\log_5 25 = 2$

EXAMPLE 118. Find $\log_{10} 0.01$.

Solution: $\log_{10} 0.01 = y$ if and only if 0.01 or $1/100 = 10^y$ but $0.01 = 10^{-2}$ so $y = -2$ and $\log_{10} 0.01 = -2$

EXAMPLE 119. Find $\log_9 27$.

Solution: $\log_9 27 = y$ if and only if $27 = 9^y$ but $27 = 9^{3/2}$ so $y = 3/2$ and $\log_9 27 = 3/2$.

EXAMPLE 120. Evaluate: a. $\log_{10}10{,}000$ b. $\log_{10}10$

c. $\log_{10}1$ d. $\log_{10}0.001$

Solution: a. $\log_{10}10{,}000 = 4$ b. $\log_{10}10 = 1$

c. $\log_{10}1 = 0$ d. $\log_{10}0.001 = -3$

EXAMPLE 121. What log expression corresponds to $125 = 5^3$?

Solution: We are given a base of 5 and an exponent of 3. A log is an exponent to an indicated base. The base is written below and between the word "log" and the number. Hence $\log_5 125 = 3$.

EXAMPLE 122. What log expression corresponds to $1{,}024 = 4^5$?

Solution: $\log_4 1{,}024 = 5$

EXAMPLE 123. What log expression corresponds to $8^{2/3} = 4$?

Solution: $\log_8 4 = \dfrac{2}{3}$.

Logarithms to the base 10 are called **common logarithms.** When one is dealing with common logs it is customary to omit the base "10". Thus $\log_{10}x$ is written $\log x$.

To understand why only positive numbers have logs assume that someone has somehow determined that $\log 0 = y$. The corresponding exponential statement is $0 = 10^y$. However, there is no exponent "y" that will make $10^y = 0$. Hence, $\log 0$ is undefined.

EXAMPLE 124. Explain why $\log(-8)$ is undefined.

Solution: $\log(-8) = y$ if and only if $-8 = 10^y$. However, there is no power of 10 that equals -8. Hence, $\log(-8)$ is undefined.

Using points from the following table we are able to graph
$y = \log_2 x$.

x	1/8	1/4	1/2	1	2	4	8
$y = \log_2 x$	-3	-2	-1	0	1	2	3

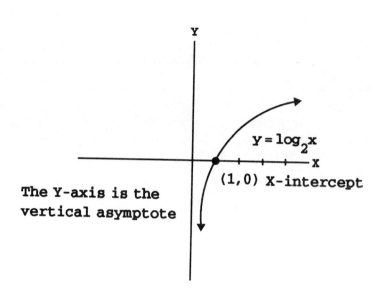

$y = \log_2 x$

(1,0) X-intercept

The Y-axis is the vertical asymptote

There are three **laws of logarithms.** Each one is based upon exponential properties. They are:

1. $\log_a xy = \log_a x + \log_a y$ The log of a product is equivalent to the sum of the logs of the numbers.

2. $\log_a \dfrac{x}{y} = \log_a x - \log_a y$ The log of a quotient is equivalent to the log of the numerator minus the log of the denominator.

3. $\log_a x^r = r \cdot \log_a x$ The log of a number to a power is equivalent to the product of the power and the log of the number.

Here are illustrations of these properties:

$\log(56)(92) = \log 56 + \log 92$

$\log \dfrac{174}{386} = \log 174 - \log 386$

$$\log 12^{39} = (39)\log 12$$

$$\log \sqrt{5} = \log 5^{1/2} = (1/2)\log 5$$

EXAMPLE 125. Assume that all variables represent positive real numbers. Use the properties of logarithms to express each of the following logs in terms of simpler logs.

    a. $\log_8 5 \cdot 12$

    b. $\log_3 \dfrac{29}{83}$

    c. $\log_a \sqrt{b}$

    d. $\log \dfrac{(x + 1)^3}{(x^2 - 1)^5}$

Solution: a. $\log_8 5 \cdot 12 = \log_8 5 + \log_8 12$

    b. $\log_3 \dfrac{29}{83} = \log_3 29 - \log_3 83$

    c. $\log_a \sqrt{b} = \log_a b^{1/2} = (1/2)\log_a b$

    d. $\log \dfrac{(x + 1)^3}{(x^2 - 1)^5} = \log(x+3)^3 - \log(x^2 - 1)^5$

$$= (3)\log(x + 1) - (5)\log(x^2 - 1)$$

EXAMPLE 126. Use the properties of logarithms to express each log as a single logarithm with a coefficient of 1. Assume all variables represent positive real numbers.

    a. $\log_3(x + 1) + \log_3(x - 1) - \log_3 7$

    b. $2 \log_a 2y - 3 \log_a y^3$

    c. $\dfrac{1}{2} \log_5 m + \dfrac{1}{3} \log_5 n^4 - \log_5 m^2 n$

Solution: a. $\log_3(x + 1) + \log_3(x - 1) - \log_3 7 =$

$$\log_3 \dfrac{(x + 1)(x - 1)}{7}$$

b. $2 \log_a 2y - 3 \log_a y^3 = \log_a (2y)^2 - \log_a y^9$

$$= \log_a \frac{4y^2}{y^9}$$

$$= \log_a \frac{4}{y^7}$$

c. $\frac{1}{2} \log_5 m + \frac{1}{3} \log_5 n^4 - \log_5 m^2 n =$

$\log_5 m^{1/2} + \log_5 n^{4/3} - \log_5 m^2 n \ =$

$$\log_5 \frac{m^{1/2} n^{4/3}}{m^2 n}$$

$$\log_5 \frac{n^{1/3}}{m^{3/2}}$$

EXAMPLE 127. Given log 2 = 0.3010 and log 3 = 0.4771, find each of the following:

    a. log 8

    b. log 6

    c. log 5

    d. log 72

    e. $\log \sqrt[3]{5}$

Solution: a. $\log 8 = \log 2^3 = 3 \log 2 = 3(0.3010) = 0.9030$

    b. $\log 6 = \log(2)(3) = \log 2 + \log 3 = 0.3010 + 0.4771$
        $= 0.7781$

    c. $\log 5 = \log \frac{10}{2} = \log 10 - \log 2 = 1 - 0.3010 \ 0$
        $= 0.6990$

    d. $\log 72 = \log(8)(9) = \log(2^3)(3^2)$

$$= \log 2^3 + \log 3^2$$

$$= 3 \log 2 + 2 \log 3$$

$$= (3)(0.3010) + 2(0.4771)$$

$$= 0.9030 + 0.9542$$

$$= 1.8572$$

e. $\log \sqrt[3]{5} = \log 5^{1/3} = (1/3) \log 5 = (1/3)(0.6990)$
        $= 0.2330$

An equation, such as $\log_8 5x = \log_8 10$, where the variable in the equation appears as part of the logarithm is called a **logarithmic equation**. Because the log expression $\log_a x$ yields a unique value for each value of x, $\log_a m = \log_a n$ if and only if $m = n$. This fact is sometimes used to solve logarithmic equations. If $\log_8 5x = \log_8 10$, then $5x = 10$ and $x = 2$. Sometimes a logarithmic equation is solved by converting the logarithmic equation to the equivalent exponential expression and then solving this equation. To solve $\log_6 2x = 3$ is equivalent to solving $2x = 6^3$ so $2x = 216$ and $x = 108$. It is important to take the time to check each "potential solution" of a logarithmic equation because a solution may be extraneous.

EXAMPLE 128. Solve $\log_2(x - 3) = 5$.

Solution: $\log_2(x - 3) = 5$ if and only if $x - 3 = 2^5$, hence:
                                $x - 3 = 32$
                                $x = 35$

Check: Substitute $x = 35$ in the original equation.

           $\log_2(35 - 3)$? 5
           $\log_2 32$? 5
                   $5 = 5$, the solution checks.

EXAMPLE 129. Solve: $\log_3 x + \log_3(x - 8) = 2$.

Solution: $\log_3 x + \log_3(x - 8) = 2$

           $\log_3(x)(x - 8) = 2$ if and only if $x(x - 8) = 3^2$,
    so
                   $x(x - 8) = 9$

             $x^2 - 8x - 9 = 0$

           $(x - 9)(x + 1) = 0$

There are two solutions to check.
Hence $x = 9$ or $x = -1$.

Check: Can x = 9?   Can x = -1?

$\log_3 9 + \log_3(9 - 8)$? 2 $\log_3(-1) + \log_3(-1 - 8)$? 2

$2 + \log_3 1$? 2   ↑   ↑

$2 + \quad 0$? 2   undefined logs

$2 \quad = 2$

x = 9 is a solution.   x = -1 is an extraneous
                              solution.

EXAMPLE 130.   Solve: $\log(x + 2) = \log(x - 7) + \log 4$.

Solution: $\log(x + 2) = \log(x - 7) + \log 4$

$\log(x + 2) = \log(4)(x - 7)$ if and only if

$x + 2 \quad = 4(x - 7)$

$x + 2 \quad = 4x - 28$

$30 \quad = 3x$

$10 \quad = x$

Check: Substitute x = 10 in the original equation.

$\log(10 + 2)$? $\log(10-7) + \log 4$

$\log 12$? $\log 3 + \log 4$

$\log 12 \quad = \log 12$, the solution checks.

Logarithms to the base "e" (approximately 2.72) are called
natural logarithms because they appear so frequently in science
and business applications.   The abbreviation for natural
logarithm is "ln".   Thus ln8 refers to the "natural logarithm of
8".

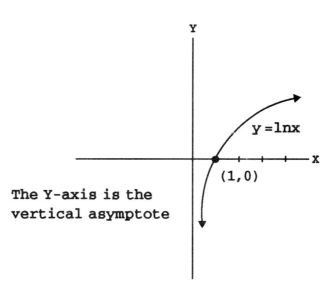

The Y-axis is the
vertical asymptote

Most natural logarithms cannot be obtained without the aid
of a table. A few noteworthy exceptions appear in the following
examples.

EXAMPLE 131.   Determine $\ln e^3$.

Solution: $\ln e^3 = y$ if and only if $e^3 = e^y$.

       This implies that $y = 3$ and $\ln e^3 = 3$.

EXAMPLE 132.   Determine $\ln e$.

Solution: $\ln e = y$ if and only if $e = e^y$.

       This implies that $y = 1$ and $\ln e = 1$.

EXAMPLE 133.   Determine $\ln 1$.

Solution: $\ln 1 = y$ if and only if $1 = e^y$.

       This implies that $y = 0$ and $\ln 1 = 0$.

2.15 TRIGONOMETRY

An **angle is in standard position** when its vertex is at the
origin, when one ray (sometimes called the "initial ray") lies
fixed upon the positive portion of the X-axis, and when the other

ray (sometimes called the "terminal ray") has been rotated counter-clockwise from the positive portion of the x-axis. All trigonometric ratios, "trig ratios", are defined in terms of an angle in standard position.

An angle is measured in **degrees** (an angle represented by a full circle has 360°) or in **radians**. The basic relationship is 180° = π radians. From this we have:

$$1° = \frac{\pi}{180} \text{ radians} = 0.017 \text{ radians}; \quad 1 \text{ radian} = \frac{180°}{\pi} = 57.32°$$

EXAMPLE 134. Convert these angle measures into degrees:

a. $\frac{\pi}{2}$ b. $\frac{3\pi}{4}$ c. 2π d. $\frac{5\pi}{6}$

Solution: a. 90°   b. 135°   c. 360°   d. 150°

EXAMPLE 135. Convert these angle measures into radians:

a. 60°   b. 72°   c. 270°   d. 315°

Solution: a. $\frac{\pi}{3}$ b. $\frac{2\pi}{5}$ c. $\frac{3\pi}{2}$ d. $\frac{7\pi}{8}$

Consider a circle of radius r > 0 centered at the origin. In this same picture the terminal ray of an angle in standard position intersects the circle at the point, P(x,y).

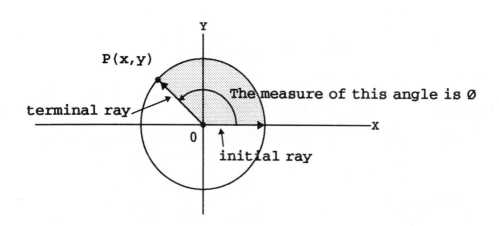

```
┌─────────────────────────────────────────────────────────────────┐
│                    Trigonometric Ratios                            │
│                                                                     │
│  Definitions                               Abbreviations           │
│                                                                     │
│  The ratio  y   is called the sine of θ          sin θ             │
│             ─                                                       │
│             r                                                       │
│  The ratio  x  is called the cosine of θ         cos θ             │
│             ─                                                       │
│             r                                                       │
│  The ratio  y  is called the tangent of θ        tan θ             │
│             ─                                                       │
│             x                                                       │
│                                                                     │
│       This ratio is undefined when x = 0 which means the           │
│       tan 90° and tan 270° are meaningless.                        │
└─────────────────────────────────────────────────────────────────┘
```

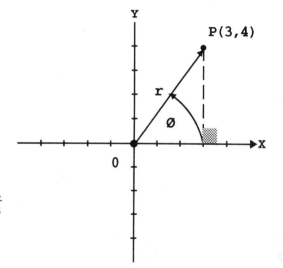

EXAMPLE 136.   Determine the sine, cosine, and tangent for the angle of measure θ to the right.

Solution: The radius r may be found using the Pythagorean Theorem:

$$r^2 = 3^2 + 4^2 \text{ so } r = \sqrt{25} \text{ or } 5$$

$$\sin \theta = \frac{4}{5} \text{ , } \cos \theta = \frac{3}{5} \text{ , } \tan \theta = \frac{4}{3}$$

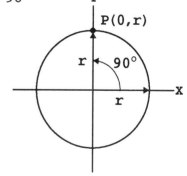

EXAMPLE 137.   Determine: a. sin 90°   b. cos 90°

Solution: Refer to the picture to the right.

a. $\sin 90° = \frac{r}{r} = 1$

b. $\cos 90° = \frac{0}{r} = 0$

The **reciprocal** of a/b where a ≠ 0 and b ≠ 0 is b/a. For example, the reciprocal of 7/8 is 8/7. In trigonometry there are the following reciprocal relationships:

The cosecant of θ, abbreviated csc θ = $\dfrac{1}{\sin \theta}$ ⎫ When the denominator

The secant of θ, abbreviated sec θ = $\dfrac{1}{\cos \theta}$ ⎬ is zero the

The cotangent of θ, abbreviated ctn θ = $\dfrac{1}{\tan \theta}$ ⎭ ratio is meaningless.

The product of two reciprocal ratios is always one. Thus (sin θ)(csc θ) = 1, (cos θ)(sec θ) = 1, and (tan θ)(ctn θ) = 1. For example, if tan θ = 2/3, then ctn θ = 3/2 and (tan θ)(ctn θ) = (2/3)(3/2) = 1.

EXAMPLE 138. cos θ = $\dfrac{9}{10}$ determine the sec θ.

Solution: sec θ = $\dfrac{1}{\cos \theta}$ = $\dfrac{1}{9/10}$ = $\dfrac{10}{9}$

If you wish to square a trigonometric ratio such as cos 5°, you may express this by writing either (cos 5°)² or cos²5°. It is wrong to write cos(5°)² because this denotes the cosine of 25°. Similarily to write "the sine cubed of 17°" write sin³17°. In general to express the fact that a trigonometric ratio is being raised to the "nth power" write "ratioⁿθ".

EXAMPLE 139. Explain the meaning of: a. cos²8°   b. cos(8°)²

Solution: a.   Here you are "squaring" the ratio.
         cos²8° = (cos 8°)(cos 8°)

      b.   Here you are "squaring" the angle.
         cos(8°)² = cos 64°

When the terminal ray of the standard position angle lies within quadrant I the sin θ, cos θ, and tan θ are positive ratios. When the terminal ray lies within quadrant II only the sin θ and its reciprocal ratio are positive, when the terminal ray lies within quadrant III, only the tan θ and its reciprocal ratio are positive, and when the terminal ray lies within quadrant IV only the cos θ and its reciprocal ratio are positive. The following phrase may help you to memorize when a trigonometric ratio is positive, "All (quadrant I) Students (the sine in II) Take (the tangent in III) Calculus (the cosine in IV).

The sine ratio and its reciprocal are positive.	All ratios are positive in this quadrant.
The tangent ratio and its reciprocal are positive.	The cosine ratio and its reciprocal are positive.

A trigonometric equation that is true for all angles for which the ratios are defined is called a **trigonometric identity**. From the relationship $x^2 + y^2 = r^2$, which is true for all angles in standard position, we obtain the following trigonometric identities:

1. $\dfrac{x^2}{r^2} + \dfrac{y^2}{r^2} = \dfrac{r^2}{r^2}$ produces $\cos^2\theta + \sin^2\theta = 1$

2. $\dfrac{x^2}{x^2} + \dfrac{y^2}{x^2} = \dfrac{r^2}{x^2}$ where $x \neq 0$, produces $1 + \tan^2\theta = \sec^2\theta$

3. $\dfrac{x^2}{y^2} + \dfrac{y^2}{y^2} = \dfrac{r^2}{y^2}$ where $y \neq 0$, produces $\text{ctn}^2\theta + 1 = \csc^2\theta$

Another important trigonometry identity is obtained as follows:

$$\tan \theta = \frac{y}{x} = \frac{y/r}{x/r} = \frac{\sin \theta}{\cos \theta} \; ; \text{ where } \cos \theta \neq 0$$

EXAMPLE 140. The terminal ray of the angle in standard position lies in quadrant I and the $\sin \theta = \sqrt{3}/2$. Use a trigonometric identity to determine $\cos \theta$.

Solution: $\cos^2\theta + \sin^2\theta = 1$

$$\cos^2\theta = 1 - \sin^2$$

$$\cos^2 = 1 - \left(\frac{\sqrt{3}}{2} \right)^2$$

$$\cos^2\theta = 1 - \frac{3}{4}$$

$$\cos^2\theta = \frac{1}{4}$$

$$\cos\theta = \sqrt{1/4} = \frac{1}{2}$$

EXAMPLE 141. If sin θ = 5/13 and cos θ = -12/13 use a trigonometric identity to determine tan θ.

Solution: $\tan\theta = \dfrac{\sin\theta}{\cos\theta} = \dfrac{5/13}{-12/13} = -\dfrac{5}{12}$

One cannot determine a trigonometric ratio for an angle with a measure of 19°, 103°, 261°, or 328° without a table. For this reason most text problems involve angles whose measure is a multiple of 30°, 45°, or 60°. This allows one to use the special 45°, 45°, 90° right triangle and the special 30°, 60°, 90° right triangle which are reproduced below for convenience.

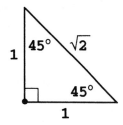 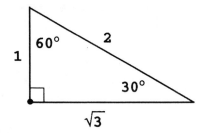

EXAMPLE 142. Determine cos 45°.

Solution: The **reference angle** to be used lies between the terminal ray and the x-axis.

$\cos 45° = \dfrac{1}{\sqrt{2}} = \dfrac{\sqrt{2}}{2}$

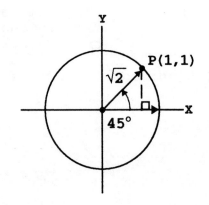

EXAMPLE 143. Determine tan 120°.

Solution: The reference angle to be used lies between the terminal ray and the x-axis.

$$\tan 120° = \frac{\sqrt{3}}{-1} = -\sqrt{3}$$

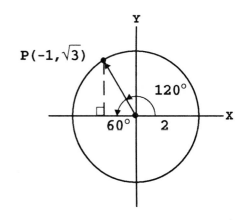

EXAMPLE 144. Determine sin 300°.

Solution: The reference angle lies between the terminal ray and the x-axis.

$$\sin 300° = \frac{-\sqrt{3}}{2}$$

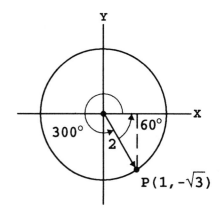

EXAMPLE 145. Determine sec 315°.

Solution: The reference angle lies between the terminal ray and the x-axis.

$\cos 315° = 1/\sqrt{2}$ so its reciprocal sec $315° = \sqrt{2}$

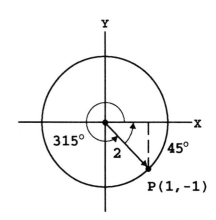

EXAMPLE 146. Determine sin(-210°).

Solution: For an angle of negative measure the rotation is opposite that of an angle of positive measure. From the picture to the right observe the terminal ray for -210° is identical to the terminal ray for 150°.

$$\sin(-210°) = \sin 150° = \frac{1}{2}$$

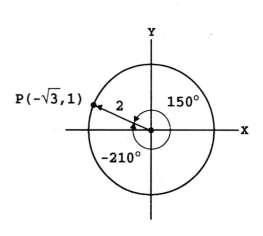

97

If the angle measure is recorded upon the x-axis, the graph of y = cos x is an infinite smooth curve that repeats every 360° (2π radians) and oscillates from a maximum distance of one unit above the x-axis to a maximum distance of one unit below the x-axis. This graph is reproduced below:

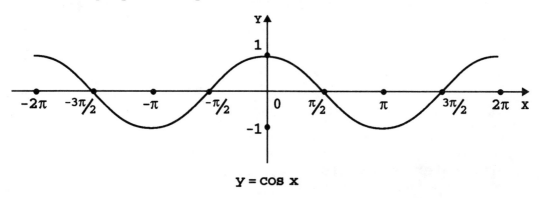

y = cos x

If the angle measure is recorded upon the x-axis, the graph of y = sin x is an infinite smooth curve that repeats every 360° (2π radians) and oscillates from a maximum distance of one unit above the x-axis to a maximum distance of one unit below the x-axis. This graph is reproduced below:

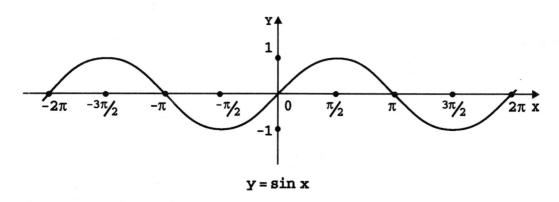

y = sin x

The graphs of y = Acos x and y = Asin x have the same form as the graphs of y = cos x and y = sin x except for a difference in the maximum distance of the graph above the x-axis. This maximum distance above the x-axis is called the **amplitude** of the graph. It is given by the number |A|.

EXAMPLE 147. What is the amplitude of y = 3 cos x?

Solution: The amplitude is |3| = 3.

EXAMPLE 148. What is the amplitude of $y = (1/2)\sin x$?

Solution: The amplitude is $|1/2| = 1/2$.

The graphs of $y = \cos BX$ and $y = \sin BX$, have the same form as the graph of $y = \cos x$ and $y = \sin x$, except for the interval of the x-axis (in degrees or radians) which is required for the graph to repeat. This interval is called **the period** of the graph. For the case that $B > 0$, the period of both $y = \cos BX$ and $y = \sin BX$ is $2\pi/B$.

EXAMPLE 149. What is the period of $y = \cos 2x$?

Solution: The period $= \dfrac{2\pi}{2} = \pi$.

EXAMPLE 150. What is the period of $y = \sin (1/3)x$?

Solution: The period is $\dfrac{2\pi}{1/3} = 6\pi$.

2.16 FUNCTIONS

It often happens in mathematics that there exists a unique formula or rule between two variables. The area of a circle, $A = \pi R^2$, depends on the value of the radius, R. When the radius equals 3, $A = 9\pi$. Because one is free to select the value of R in this formula, R is called the **independent variable**. The value of A depends on the value assigned to R, and hence A is called the **dependent variable**.

> A **function**, commonly denoted "f", is a rule that associates to each value of the independent variable a unique value of the dependent variable.

The relationship $y = \pm \sqrt{x}$, is not a function because for each positive value of x it is requesting both square roots.

99

When x = 9, y is both +3 and -3. In order for a formula to be a function, each value of the independent variable must produce a single value of the dependent variable.

A real number assigned to the independent variable is called a domain-value. The set of all domain-values is called the **domain** of the function. In our illustration (A = πR²) R = 5 is a domain-value. Some other domain-values are: R = 3/4, 2, 5.60, 7.99, and 20. For each domain-value the dependent variable takes on a unique value called a range-value. The set of all range-values is called the **range**. When R = 5, the range-value is 25π.

The formula F = (9/5)C + 32 produces a unique Fahrenheit temperature, F, for each Centigrade temperature, C. Here, the independent variable is C and the dependent variable is F. If C is assigned the domain-value 5°, then F assumes the range-value 41°.

Whenever it is reasonable to do so, the dependent variable is either denoted "y" or "f(x)"; and the independent variable is denoted "x". This convention allows us to visualize the domain as part or all of the x-axis that is so familiar to all of us. It allows us to record range values as vertical measures on our graph or equivalently as y-values on our graph. All of this means that a functional relationship between two variables will usually be graphed on the familiar x-axis and y-axis of algebra. Some examples of functions in the independent variable x are:

$$f(x) = x^2, \quad f(x) = \frac{x + 1}{x + 2}, \quad x \neq -2, \quad \text{and} \quad f(x) = \sqrt{x - 4}, \quad x \geq 4$$

For the function f(x) = x², the domain-value x = 2 is recorded on the x-axis, and it's range-value, which is 4, is recorded vertically on the y-axis.

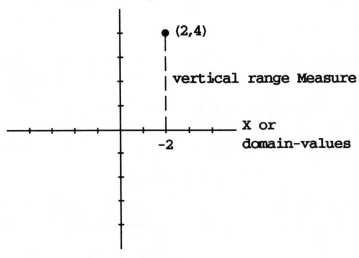

Y or f(x) or range-values

100

To determine the domain of a function begin by assuming that it is the set of Reals (Geometrically the real numbers correspond to the points on the x-axis.) and delete:

1) all values of the independent variable that demand a division by zero because division by zero is undefined.

2) all values of the indpendent variable that make the radicand of a square root or an even-root negative because such a root is not a real number.

3) all values of the independent variable that don't fit a constraint of an applied problem.

EXAMPLE 151. Determine the domain of the function $f(x) = x^2$.

Solution: There is no value of x that demands a division by zero. Because there is no radical in this formula there cannot be an even-radical with a negative radicand. There are no constraints on the variable x.

Answer: The domain is the set of reals.

EXAMPLE 152. Determine the domain for the function $f(x) = \dfrac{x + 1}{x + 2}$

Solution: The value $x = -2$ must be omitted because it will demand a division by zero.

Answer: The domain is the set of all reals except $x = -2$.

EXAMPLE 153. Determine the domain of the function $f(x) = \sqrt{x - 4}$

Solution: x cannot assume the value -10, -2, 0, 3, or any value less than 4 because it will make the radicand negative and the formula will not produce a real number.

Answer: The domain is $x \geq 4$.

EXAMPLE 154. Determine the domain of the function $A = f(R) = \pi R^2$

Solution: For all circles (We assume that a point is not a circle.) the radius measure must be a positive number.

Answer: The domain is R > 0.

     A function doesn't have to be named "f". When one has more than one function it may be convenient to name the function "g" or "h" or so on. The independent variable doesn't have to be named "x". An economist may decide to call the independent variable "i" for interest. A physicist may call the independent variable "t" for time. The function $C = 2\pi R$ relating the circumference of a circle to its radius calls the dependent variable C and the independent variable R.

     If one wishes to replace x by 2 in the function $f(x) = x^3 - 5$, this is written f(evaluated when x = 2) or simply $f(2)$. $f(2) = 2^3 - 5 = 3$. Similarly $f(-1)$ means to replace x by -1 in the function f(x). This is sometimes expressed "evaluate the function at x = -1.". $f(-1) = (-1)^3 - 5 = -6$.

EXAMPLE 155. The area of any square, with a side of measure s, is $A = f(x) = s^2$.

     a.   What is the independent variable?

     b.   What is the dependent variable?

     c.   What is the domain?

     d.   If s = 1/3 inch, what is f(1/3)?

Solution: a.   s

         b.   A

         c.   s > 0

         d.   1/9 square inch

EXAMPLE 156.

    a.   For $f(x) = x^4$, determine $f(-2)$.

    b.   For $g(x) = 5x^2 - x + 10$, determine $g(-1)$.

    c.   For $h(x) = \dfrac{3x + 5}{x^2 - 1}$, determine $h(3)$.

    d.   For $k(x) = \sqrt{x - 1}$, determine $k(17)$.

Solution: a. $f(-2) = (-2)^4 = 16$

    b. $g(-1) = 5(-1)^2 - (-1) + 10 = 5 + 1 + 10 = 16$

    c. $h(3) = \dfrac{3(3) + 5}{3^2 - 1} = \dfrac{14}{8} = 1\,\dfrac{3}{4}$

    d. $k(17) = \sqrt{17 - 1} = \sqrt{16} = 4$

EXAMPLE 157.   In a certain state the selling price of an article, S, is the purchase price, P, plus a sales tax of 5% of the purchase price on any article that costs $10 or more.

    a.   Express the selling price as a function of the purchase price.

    b.   What range-value is produced by a domian-value of $80?

    c.   What is the domain of this function?

    d.   What is the range of this function?

Solution: a. $S = f(P) = P + 0.05P$

    b. $f(\$80) = \$80 + (0.05)(\$80) = \$84$

    c. $P \geq \$10$

    d. $S \geq \$10.50$

    The graph of a relationship is a function when every vertical line intersects the graph in at most one point. This is sometimes called "the vertical line test of a function". When a graph is given, this is the fastest method of determining whether or not it represents a function.

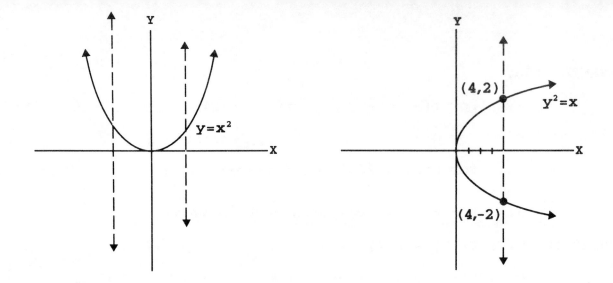

In the illustration above and to the left every vertical
line intersects the graph of $y = x^2$ in one point and hence, this
is the graph of a function of x. By contrast, a vertical line
may intersect the graph of $y^2 = x$ in two points. The graph of
$y^2 = x$ is not the graph of a function of x. Here a value of x
usually doesn't produce a unique y-value.

A beginner determines the range of a function by looking at
its graph. Because range-values are recorded vertically on the
y-axis the y-values of points on the graph of the function
represent the range of the function. What is the domain and the
range of the function below?

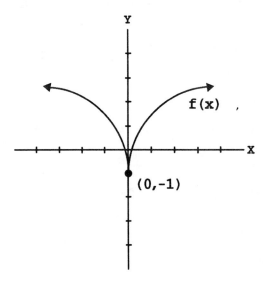

The x-values of points on this graph tell us that the domain
is the set of real numbers. The y-values of points on this graph
tell us that the range is $y \geq -1$.

104

EXAMPLE 158. Discuss the domain and the range of $f(x) = |x|$.

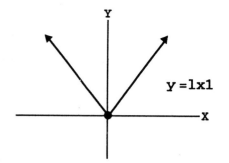

Solution: The domain is the set of real numbers. The range is $y \geq 0$.

EXAMPLE 159. Discuss the domain and the range of $f(x) = e^x$.

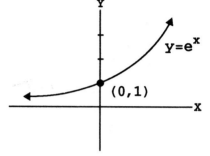

Solution: The domain is the set of real numbers. The range is $y > 0$.

EXAMPLE 160. Discuss the domain and the range of $f(x) = \ln x$.

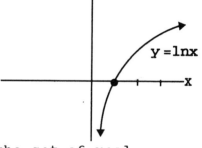

Solution: The domain is $x > 0$. The range is the set of real numbers.

EXAMPLE 161. Discuss the domain and the range of the graph of the function $f(x)$ to the right.

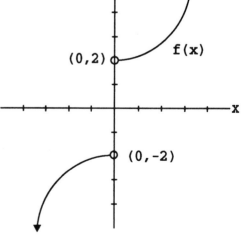

Solution: The domain is all real numbers except $x = 0$. The range is $y < -2$ or $y > 2$.

When the graph of your function is represented by a straight line that is not horizontal the domain is the set of real numbers and the range is the set of real numbers because every real number will correspond to some point on your graph. When your graph represents a polynomial function of the form
$g(x) = a_n x^n + a_{n-1} x^{n-1} + \ldots a_2 x^2 + a_1 x + a_0$, one must sketch the graph before attempting to state its range.

EXAMPLE 162.

    a.   Graph the polynomial function $g(x) = x^3 - 3x - 2$.

    b.   State its domain.

    c.   State its range.

Solution: a.

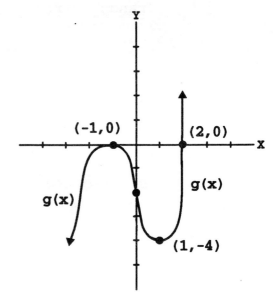

    b.   The real numbers.

    c.   The real numbers.

The **greatest integer function**, some people call it the **bracket function** because it is denoted $f(x) = [x]$, is defined for all real numbers. The function associates to any real number x a range-value that is the greatest integer less than or equal to x. For example $f(5) = [5] = 5$, $f(4.7) = [4.7] = 4$, $f(-2.05) = [-2.05] = -3$, and $f(0) = [0] = 0$.

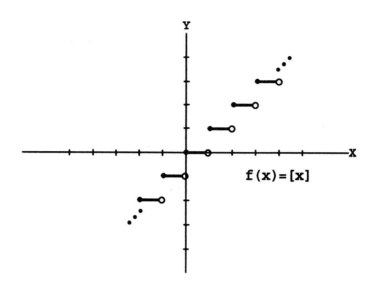

f(x)=[x]

The graph of the greatest integer function looks like a steadily climbing set of steps moving upward left to right. By looking at the y-values of the points on this graph, one determines that the range of this function is the set of integers.

EXAMPLE 163.   Determine:

        a.   [9]

        b.   [5 1/6]

        c.   [-7.83]

Solution: a.   9

        b.   5

        c.   -8

2.17 PROBLEM SOLVING

An angle is 8 times as large as its complement. Determine the degree measures of the angles. In order to solve a word problem one must understand all of the words in the problem. Two angles are complementary when their sum is 90°. Algebra skills are often helpful in solving a word problem. Call the measure of the smaller angle x and the measure of the larger angle 8x. This means that x + 8x = 90° and x = 10°. The angle measures are 10° and 80°.

A student has an average grade of 86.5 on four exams. What is the total of her exam grades? Sometimes thinking about simpler numbers will help you to solve your word problem. If your average grade on two exams was 70, then the total of the two exams must have been (2)(70) or 140. This means the total of the four exam grades that we are seeking must be (4)(86.5) = 346.

TO SOLVE A WORD PROBLEM

1. Understand the problem:

    Read the problem carefully. Do you understand all of the words in the problem? Do you understand the information that is being given? Do you understand the information that is being asked for? At this point, you are in good shape if you can state the problem in your own words.

2. Analyze the problem:

    If it helps draw a picture, a table, or a diagram. Try a similar problem with easier numbers. Can you express this problem in algebra? Is it an equation? Is it an inequality? Is the problem similar to one that you have already solved, or to one that the text has already solved? At this point, don't be afraid to use the method of "trial and error".

3. Try to solve the problem:

    Don't be afraid to try to solve a problem a second or a third time. Many problems are not solved on the first try. If you reach the point where you feel unable to solve the problem:

(1) See if the text has an answer to your problem. The answer may serve as a hint.

(2) See if there is a text example that is similar to your problem.

(3) See if a preceding problem is similar to your present problem.

4.   Check your solution:

Does your solution satisfy the conditions set forth in the problem? Is your solution reasonable? If you estimate values in the problem, does your solution compare favorably to the estimated solution?

EXAMPLE 164.   The sum of three consecutive even numbers is 222. Find the three numbers.

Solution: We begin with an estimate of the answer. If the smallest even number is 50, the sum of three consecutive even numbers is too small. If the smallest even number is 100, the sum of three consecutive even numbers is too large. The smallest even number that we are seeking must be between 50 and 100. Because each consecutive even number is two more than the preceding number, our numbers must be x, x + 2, and x + 4. We are given that the sum of the numbers is 222 so ...

$$x + (x + 2) + (x + 4) = 222$$

$$3x + 6 = 222$$

$$3x = 216$$

$$x = 72$$

Answer: 72, 74, and 76.

Check: 72 + 74 + 76 = 222

EXAMPLE 165.   Twenty-one people are at a party. If every female at the party is dancing with a male partner and there are three males remaining without partners, how many females are at the party?

Solution: Using algebra allows us to express information in a succinct manner. If "x" represents the number of females, then "x + 3" represents the number of males. But there are 21 people in all so we can write this equation:

$$x + (x + 3) = 21$$

$$2x + 3 = 21$$

$$x = 9$$

There are 9 females and 12 males.

Check: 9 + 12 = 21

EXAMPLE 166.   One side of a triangle is 3 inches long and the second side of the triangle is 5 inches long. What length is possible for the third side?

Solution: Let's begin with a little sketch of the problem.

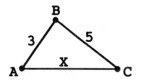

In our sketch the length of the third side is x inches. Recall that the sum of the lengths of any two sides of a triangle must exceed the length of the remaining side. So:

$$3 + x > 5 \quad \text{and} \quad 3 + 5 > x$$

$$x > 2 \qquad\qquad 8 > x$$

The answer is 2 < x < 8.

EXAMPLE 167.   A woman has only nickels and dimes in her change. She has 11 more nickels than dimes. If she has a total of $2.65 in change, how many of each type of coin does she have?

Solution: Thinking of $2.65 as 265 pennies will eliminate decimals. If x represents the number of dimes, then x + 11 will represent the number of nickels.

number of coin	value in pennies
x dimes	10x
x + 11 nickels	5(x + 11)

We are given that our nickels and dimes total 265 pennies so ...

$$10x + 5(x + 11) = 265$$

$$10x + 5x + 55 = 265$$

$$15x = 210$$

$$x = 14$$

Answer: 25 nickels, 14 dimes

Check: Does our answer satisfy the conditions given in the problem? 25 nickels is $1.25 and 14 dimes is $1.40. Because $1.25 + $1.40 = 2.65 and there are exactly 11 more nickels than dimes the answer is verified.

EXAMPLE 168.   How many chords can be drawn using four distinct points A, B, C, and D on the circumference of a circle?

Solution: Recall a chord is a straight line segment joining two points on the circumference of a circle. Here a sketch may help. From A one can draw AB, AC, and AD.

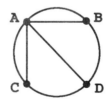

From B one can draw two additional chords BC and BD (chord BA is the same as chord AB). One final chord can be drawn between C and D. Six chords can be drawn.

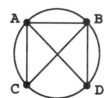 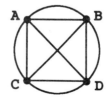

EXAMPLE 169. Two men A and B travel toward each other (in a straight line) from points 440 miles apart. If A is traveling 50 miles per hour and B is traveling 60 miles per hour, in how many hours will they meet?

Solution: Here it may be helpful to make a table listing the parts of the formula d = rt.

	rate of travel	time to meeting	= distance
for A	50 miles/hr	t hours	50t
for B	60 miles/hr	t hours	60t

Because they meet after t hours we have this equation:

$$50t + 60t = 440$$

$$110t = 440$$

$$t = 4 \text{ hours}$$

EXAMPLE 170. Attendance at the first two performances of a class play was 185 and 205. At least how many people must attend the third performance if the average of the three is to be at least 200 people?

Solution: We are being asked to obtain an "at least" number or a minimum number of people who must attend the third performance. Call the minimum number x. Remembering that an average of three numbers is the sum of the numbers divided by three we have:

$$\frac{185 + 205 + x}{3} \geq 200$$

$$185 + 205 + x \geq 600$$

$$390 + x \geq 600$$

$$x \geq 210$$

Answer: We will obtain the desired average if at least 210 people attend.

Check: $\frac{185 + 205 + 210}{3} = 200$

112

EXAMPLE 171. If a baseball player hits 10 home runs in the first 45 games of the season and one assumes that he can continue at this pace, how many home runs can be expected at the end of a 162 game season?

Solution: Let x be the number of home runs in 162 games.

$$\frac{x \text{ home runs}}{162 \text{ games}} = \frac{10 \text{ home runs}}{45 \text{ games}}$$

$$45x = (10)(162)$$

$$x = 36$$

Check: Intuitively, this answer seems reasonable because 162 games is somewhat less than 4 times 45 games so one expects the answer to be somewhat less than (4)(10) = 40.

EXAMPLE 172. In triangle ABC angle A has a measure of 70° and angle B has a measure of 50°. What is the measure of an exterior angle located at vertex C?

Solution: In order to solve a word problem you need knowledge of the subject matter with which you are dealing. Maybe you remember a geometry theorem that states, "The measure of an exterior angle of a triangle is equal to the sum of the measures of the two interior nonadjacent angles." If you remember this theorem you are done. If not a sketch may help.

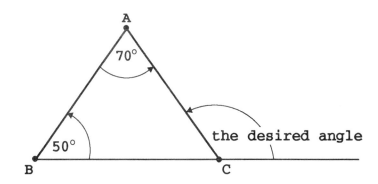

We all remember that the sum of the angle measures of a triangle is 180°. Hence, angle C must contain 60°, and because we want the measure of the supplement of this angle, the answer must be 120°.

EXAMPLE 173. Determine the simple interest, I, and the amount, A, on a loan of $20,000 at rate of 9% for 72 days.

Solution: Recall, the formula for simple interest is I = PRT where P is the principal, R is the rate, and T is the time in years. The amount at the end of the time period is given by the formula A = P + I or P(I + RT). Realizing that 72 days is 1/5 or 0.20 of a year we have:

$$I = (\$20,000)(0.09)(0.20)$$

$$= \$360 \quad \text{and}$$

$$A = \$20,000 + \$360 = \$20,360$$

EXAMPLE 174. Gleason Peanut Co. wishes to mix $0.85 peanuts with $1.15 cashews in a 12-pound container. If the mixture is to sell at $1.10 per pound, how many pounds of each must be mixed in a 12-pound container?

Solution: Here is a table representing the information of this problem.

type of nut	amount	unit cost	=	value
peanuts	x	$0.85		0.85x
cashews	12 - x	$1.15		1.15(12 - x)
blend	12	$1.10		1.10(12)

Now we are going to write algebraically what we have been told.

$$0.85x + 1.15(12 - x) = 1.10(12)$$

$$100[0.85x + 1.15(12 - x)] = (100)(1.10)(12)$$

$$85x + 115(12 - x) = 110(12)$$

$$-30x = -60$$

$$x = 2$$

Answer: Each 12-pound container must contain 2 pounds of peanuts and 10 pounds of cashews.

EXAMPLE 175. Flying against a head wind an airplane can fly 3,000 kilometers in 6 hours. At the same air speed it can make the return flight with the wind in 5 hours. Find the air speed of the airplane and the wind speed.

Solution: To begin observe that we are being asked to find both the speed of the airplane, call it "x", and the wind speed, call it "y". Here a table recording the given information will help.

	ground speed, r	time, t	= distance
with head wind	x - y	6 hours	6(x - y)
with tail wind	x + y	5 hours	5(x + y)

Now let's try to write equations to express the given information.

1.   The distance with a head wind is 3,000 kilometers.

$$6(x - y) = 3,000$$

2.   The distance with a tail wind is 3,000 kilometers.

$$5(x + y) = 3,000$$

At this point we need to solve the following system of two equations in two variables:

1. $6(x - y) = 3,000$ 1. $x - y = 500$

and

2. $5(x + y) = 3,000$ 2. $x + y = 600$

Answer: x or the air speed of the airplane is 550 km/hr and y or the wind speed is 50 km/hr.

Check: 1. $6(550 - 50) = 6(500) = 3,000$

2. $5(550 + 50) = 5(600) = 3,000$

INDEX

M

means of a proportion, 47
minuend, 44
mixed number, 39
multiplicand, 45
multiplier, 45
multiplying
  decimals, 45
  fractions, 41

N

natural exponential equation, 82
natural logarithm, 90
natural number, 38
negative square root, 72
nonterminating decimal, 43
numerator, 38

P

parabola, 79
period of a trigonometric ratio, 99
Pi (π), 63
positive square root, 72
power, 49
principal square root, 72
Principle of a Zero Product, 75
Product Rule for Radicals, 73
proper fraction, 38
proportion, 47

Q

quadratic equation, 75
quadratic formula, 77
quotient, 46

R

radian, 92
radical sign, 72
radicand, 72
range, 100
ratio, 46
reciprocal, 93
reference angle, 96
remainder, 46
root, 57
Rule for Proportions, 47

S

slope, 66
slope-intercept form of line, 68
solution, 57
standard form of a quadratic, 77
subtraction
  decimals, 44
  fractions, 40
subtrahend, 44

T

terms of a proportion, 47
terminal ray, 91
terminating decimal, 43
total, 44
trigonometric identity, 95
trigonometric ratio, 93

V

vertex, 79

W

whole number, 38
word problems, 108

X

x-intercept, 70

Y

y-intercept, 70

STUDENT NOTES

STUDENT NOTES

STUDENT NOTES

STUDENT NOTES

STUDENT NOTES